CE MARKING FOR MACHINERY DIRECTIVE

CE MARKING FOR MACHINERY DIRECTIVE

Authored by SWBC International

NEW YORK   **ASME PRESS**   **1999**

© 1999 by The American Society of Mechanical Engineers
Three Park Avenue, New York, NY 10016

All rights reserved. Printed in the United States of America. Except as permitted under the United States Copyright Act of 1976, no part of this publication may be reproduced or distributed in any form or by any means, or stored in a database or retrieval system, without the prior written permission of the publisher.

Information contained in this work has been obtained by the American Society of Mechanical Engineers from sources believed to be reliable. However, neither ASME nor its authors or editors guarantee the accuracy or completeness of any information published in this work. Neither ASME nor its authors and editors shall be responsible for any errors, omissions, or damages arising out of the use of this information. The work is published with the understanding that ASME and its authors and editors are supplying information but are not attempting to render engineering or other professional services. If such engineering or professional services are required, the assistance of an appropriate professional should be sought.

ASME *shall not be responsible for statements or opinions advanced in papers or . . . printed in its publications* (B7.1.3). Statement from the Bylaws.

Authorization to photocopy material for internal or personal use under circumstances not falling within the fair use provisions of the Copyright Act is granted by ASME to libraries and other users registered with Copyright Clearance Center (CCC) Transactional Reporting Service provided the base fee of $4.00 per page is paid directly to the CCC, 222 Rosewood Drive, Danvers, MA 01923.

Library of Congress Cataloging-in-Publication Data

    CE marking for machinery directive / SWBC International
        p.   cm.
    Includes bibliographical references and index.
    ISBN 0-7918-0091-1
    1. Machinery industry—Law and legislation—European Union countries.
    2. Machinery—Safety regulations—European Union countries. I. SWBC
      International
KJE6763.8.C4   1999
341.7'54—dc21

                                                    99-33668
                                                       CIP

Table of Contents

PREFACE   **IX**

I. CE IN GENERAL   **1**

1.1 The history of the European Union 3
1.2 The Rules and Regulations of the European Union 6
    1.2.1 Regulations 6
    1.2.2 Directives 8
    1.2.3 Decisions 8
    1.2.4 Recommendations and Advice 9
1.3 European legislation for products:
    New Approach versus Old Approach 9
    1.3.1 Old Approach 9
    1.3.2 New Approach 11
    1.3.3 Advantages and limitations of the New Approach 12
1.4 Standards and harmonization 15
    1.4.1 Definition 15
    1.4.2 How are standards drafted? 16
    1.4.3 European standards and harmonization 17
1.5 Conformity-Assessment procedures 19
    1.5.1 Inspection and certification by the importer 21
    1.5.2 Inspection by a Notified Body 21
    1.5.3 Verifying observance by the government 23

1.6 Quality systems and CE Marking 24
1.7 Consequences for the American manufacturer 27
    1.7.1 Introduction. 27
    1.7.2 Responsibilities 28
    1.7.3 Steps for exporting from the USA to the EEA. 29
    1.7.4 Mutual Recognition Agreements (MRAs) 31
1.8 Summary of New Approach Directives and proposals. 32
    1.8.1 Overview of all New Approach Directives 32
        1.8.1.1 Active implantable medical devices—90/385/EEC 32
        1.8.1.2 Equipment and protective systems intended
               for use in a potentially explosive environment—
               94/009/EEC. 33
        1.8.1.3 Simple pressure vessels—87/404/EEC
               and 90/488/EEC 34
        1.8.1.4 Electromagnetic compatibility (EMC) 89/336/EEC
               and 92/31/EEC 34
        1.8.1.5 Appliances burning gaseous fuels—90/396/EEC. 35
        1.8.1.6 Low voltage—73/23/EEC 35
        1.8.1.7 Machinery—89/392/EEC, 91/386/EEC
               and 93/44/EEC 36
        1.8.1.8 Medical devices—93/42/EEC. 37
        1.8.1.9 Non-automatic weighing machines—90/384/EEC 37
        1.8.1.10 Personal protective equipment—89/686/EEC. 37
        1.8.1.11 Telecommunications terminal equipment
                and satellite earth station equipment—98/13/EC 38
        1.8.1.12 Efficiency requirements for new hot-water boilers
                fired with liquid or gaseous fuels—92/42/EEC. 38
        1.8.1.13 Safety of toys—88/378/EEC 39
        1.8.1.14 Domestic electric refrigerators,
                deep freezers—96/57/EC. 39
        1.8.1.15 Building materials—89/106/EEC 40

    1.8.1.16 Pleasure craft—94/25/EEC 40

    1.8.1.17 Explosives for civil use—93/15/EEC 41

    1.8.1.18 Pressure equipment—97/23/EC. 41

    1.8.1.19 Lifts—95/16/EC. 42

    1.8.1.20 High-speed trains—96/48/EC 42

    1.8.1.21 Measuring equipment—proposal 42

    1.8.1.22 Cableway installations for personnel
               transport—proposal 42

    1.8.1.23 Precious metals—proposal 43

    1.8.1.24 In vitro diagnostics—proposal 43

  1.8.2 The CE Marking Directive 44

2. MACHINERY DIRECTIVE—GENERAL   47

2.1 The Directives (89/392/EEC, 91/386/EEC, 93/44/EEC, and
    93/68/EEC) 49

2.2 Area of application 49

2.3 Exceptions 52

2.4 Machines with an increased risk 54

2.5 Relations with other Directives. 55

2.6 Standards 56

  2.6.1 Type A standards 56

  2.6.2 Type B standards 58

  2.6.3 Type C standards 60

2.7 Responsibilities. 61

  2.7.1 Manufacturer of complete machines. 63

  2.7.2 Manufacturer of machine parts
       and semi-finished products 65

  2.7.3 Authorized representative of the manufacturer. 65

  2.7.4 Importer of machines 66

2.8 Assembly of semi-finished products 67
2.9 Private label 68
2.10 Installation of machines 68
2.11 Hiring of machines 70
2.12 The sale of machines with a CE Marking outside the EEA 71

3. PROCEDURE FOR CE MARKING UNDER THE MACHINERY DIRECTIVE   73

3.1 Introduction 75
3.2 Risk assessment 75
   3.2.1 Introduction. 75
   3.2.2 Definitions 76
   3.2.3 Step plan 77
   3.2.4 Identification of risks. 77
   3.2.5 Assessment of the risks 78
   3.2.6 Evaluation of the risks. 79
   3.2.7 Risk reduction 79
   3.2.8 Safety evaluation 80
3.3 Technical Construction File 80
3.4 The User's Manual 81
   3.4.1 Introduction. 81
   3.4.2 The language of the user's manual 82
   3.4.3 Structure of the user's manual. 82
   3.4.4 Method and tools. 83
   3.4.5 Technical specification and other technical data. 84
   3.4.6 Safety aspects 85
   3.4.7 Transport, installation, taking into operation,
         and dismantling. 86
   3.4.8 Instructions for use 89

3.4.9 Maintenance and repairs..........................89
  3.4.10 Environment 90
3.5 EC Type Declaration......................................91
3.6 EC Declarations according to Annex II
    of the Machinery Directive...............................91
  3.6.1 EC Declaration of Conformity IIA...................92
  3.6.2 EC Declaration of Conformity IIB...................93
  3.6.3 EC Declaration of Conformity IIC...................95
3.7 Affixing the CE Marking..................................96

4. PRACTICAL EXAMPLES OF THE MACHINERY DIRECTIVE   97

4.1 Introduction...99
4.2 The most frequently asked questions
    on the Machinery Directive...............................99
4.3 Practical example of risk assessment....................109
  4.3.1 Introduction......................................109
  4.3.2 Step 1: risk identification........................109
    4.3.2.1 Identification of the machine..................109
    4.3.2.2 Identification of users........................109
    4.3.2.3 Identification of conditions of use.............110
  4.3.3 Relationship with the Machinery Directive..........112
  4.3.4 Specific requirements and standards................113
    4.3.4.1 Specific requirements..........................113
    4.3.4.2 Specific standards.............................114
  4.3.5 Step 2: risk assessment 115
    4.3.5.1 Product process phases and their purpose.......115
    4.3.5.2 Description of product process phases for the hedge
            clippers 116

    4.3.6 Step 3: risk evaluation 116
    4.3.7 Step 4: risk reduction........................... 116
        4.3.7.1 Intrinsically safe design 116
        4.3.7.2 Taking protective measures 117
        4.3.7.3 Providing information to the user 117
    4.3.8 Step 5: safety evaluation 119
4.4 EC Declaration of Conformity 121
4.5 Affixing the CE Marking................................ 121

5. LIABILITY WITHIN THE EEA   123

5.1 Introduction 125
5.2 What is liability?................................... 125
    5.2.1 Definition..................................... 125
    5.2.2 Who can be held liable?......................... 125
    5.2.3 When is someone liable? 126
    5.2.4 Proving liability 126
    5.2.5 Exclusion of liability 127
5.3 Relationship between CE Marking and product liability 127
    5.3.1 CE Marking—product liability..................... 127
    5.3.2 No complete exclusion of liability 127
    5.3.3 The advantage of CE Marking in relation to liability 128
    5.3.4 Establishing contractual matters.................. 129
    5.3.5 Insurance..................................... 129
    5.3.6 Criminal liability 130
5.4 Preventive measures: Step plan 130
5.5 Approach in the event of damage and claims:
    Check list for the end producer 134

APPENDIX I: MACHINERY DIRECTIVE (89/392/EEC, 91/368/EEC AND 93/44/EEC)	**139**
APPENDIX II: LIST OF STANDARDS	**229**
APPENDIX III: CHECKLIST	**243**
APPENDIX IV: ADDRESSES WITH REGARD TO THE MACHINERY DIRECTIVE	**253**
APPENDIX V: INTEGRATED MACHINERY DIRECTIVE (98/37/EC)	**269**
GLOSSARY	**349**
QUOTATION OF SOURCES	**361**
INDEX	**363**

Preface

Effective 1 January 1995, machines, installations, and safety components fall under a European Directive, namely, the Machinery Directive. From the date cited, these products must comply with European safety regulations. Machines, installations, and safety components which do not comply with these safety requirements may not be sold or used in the European Economic Area (EEA). This stipulation brings about many consequences, not only for manufacturers and importers within the EEA but also for manufacturers and exporters outside the EEA.

From the introduction of the Machinery Directive, interest in the matter has grown rapidly. From industry, in particular, there were many questions on how it was possible to make machines, installations, and safety components comply with the safety requirements arising from the Machinery Directive.

This book gives a clear and complete understanding of the legislation and answers questions that arise from practical situations in which the Machinery Directive plays a role. It has been specifically written for the American market and is intended for anyone dealing with machines, installations, and safety components professionally or indirectly.

You will gain a clear understanding of the requirements imposed by the Machinery Directive. A step-by-step explanation of the certification procedure is given. You are provided with support in the form of clear texts, practical examples, well-organized diagrams, and drawings.

Because practice always goes hand in hand with legislation, the complete text of the Machinery Directive is included in Appendix I. You will also find an overview of the European harmonized standards which are applicable. You could not wish for a more complete and practical reference work. We hope this book clarifies matters for anyone who has anything to do with the Machinery Directive.

Please note that Machinery Directive 89/392/EEC and amendments 91/386/EEC (only Article 1), 93/44/EEC, and 93/68/EEC (only Article 6) were repealed in August 1998 and replaced by Directive 98/37/EC. This Directive integrates the previous Directive 89/392/EEC and its amendments and has been incorporated in Appendix V of this book.

SWBC International

1
CE in General

1.1 THE HISTORY OF THE EUROPEAN UNION

The European Community (EC) had its origins in 1951, when the European Coal and Steel Community (ECSC) treaty was signed. Additional treaties were signed in Rome in 1957; in all, the European Community was based on three separate treaties:

- European Coal and Steel Community (ECSC), 1951
- European Economic Community (EEC), 1957
- European Atomic Energy Commission (Euratom), 1957

One of the objectives of the European Community is the realization of the internal market. This means that a single market is to be created without any trade restrictions. The single market is characterized by what are known as the "four freedoms":

1. Free movement of goods
2. Free movement of persons
3. Free movement of services
4. Free movement of capital

Since the EEC began, attempts have been made to achieve an internal market by harmonizing legislation among Member States. The internal market comprises an area without internal borders. Within this area, the free movement of goods, persons, services, and capital are safeguarded according to the conditions of the EEC treaty.

With the internal market, an attempt has been made to create the conditions for free trade between the Member States, which will mean that fiscal and technical trade restrictions will disappear. By harmonizing the legislation there will be no more technical trade restrictions. The European Union (EU) currently comprises fifteen Member States: Austria, Belgium, Denmark, Finland, France, Germany, Greece, Ireland, Italy, Luxembourg, the Netherlands, Portugal, Spain, Sweden, and the United Kingdom. (See the map in Figure 1.1.)

FIGURE 1.1 The countries of the European Union and the European Economic Area

Article 30 of the EEC treaty talks about the free movement of goods. This Article states that "quantitative import restrictions and all measures of similar effect between the Member States are prohibited." The Article gives rise to a free movement of goods. Article 36 of the EEC treaty presents an exception to this, stating that "if goods are dangerous to the safety of persons or to the environment, then Member States are free to exclude them from their territory." This also explains the introduction of the New Approach Directives, discussed in Section 1.3 of this chapter. By appealing to Article 36 of the EEC treaty, Member States will be able to exclude goods from their territory due to certain laws containing safety requirements. This means that a technical trade barrier is in fact put in place.

Each Member State was permitted to set its own safety requirements. This meant it was quite feasible that a Dutch manufacturer exporting his products to, for example, France, would have to comply with different safety requirements than in the Netherlands. This, however, clearly contradicts the EU's intended free-trade objectives. This is why the EU has decided to harmonize the different national legislation of the Member States, through the New Approach Directives. Products that comply with the requirements of the New Approach Directives are labeled with the letters "CE" (see Figure 1.2); the label must be attached to the product in a permanent manner, as explained at the end of Chapter 4.

CE stands for "Conformité Européenne."

This label means that the product complies with the requirements of all the obliged New Approach Directive(s). Translated literally, Conformité Européenne means "European Conformity."

The treaty of the European Economic Area (EEA) came into force in July 1993. The fifteen countries of the European Union as well as

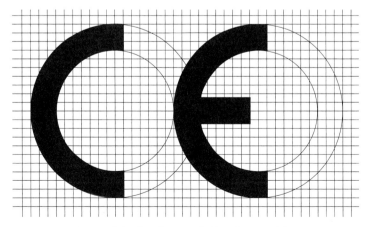

FIGURE 1.2 The form of the CE Marking

Iceland, Norway, and Liechtenstein belong to the EEA. This means that the European Directives also apply in these countries. Switzerland is not one of the countries of the EU, nor does it belong to the EEA. Switzerland should therefore be seen as a country situated, as it were, outside the European Union.

1.2 THE RULES AND REGULATIONS OF THE EUROPEAN UNION

The rules and Regulations of the EU are expressed in four ways as illustrated in Figure 1.3, via:

1. Regulations
2. Directives
3. Decisions
4. Recommendations and Opinions

It is self-evident that the Rules and Regulations of the EU are established with the qualified unity of the Member States. However, a consequence of this method of decision-making is that some time may pass before agreement is reached.

Everything established as binding in the European Rules and Regulations has priority over national legislation. A Member State may never act contrary to a European Regulation, Directive, or Decision.

1.2.1 Regulations

Regulations have a general tenor and all their components are binding and are directly applicable to all Member States. A general tenor means that a Regulation is in principle applicable to an indeterminate number of cases and persons.

The direct applicability of a Regulation means that it does not first have to be transposed into national legislation. This direct effect of Regulations also means that a Regulation can directly provide rights to private individuals within the Member States. The national courts are obliged to uphold these rights.

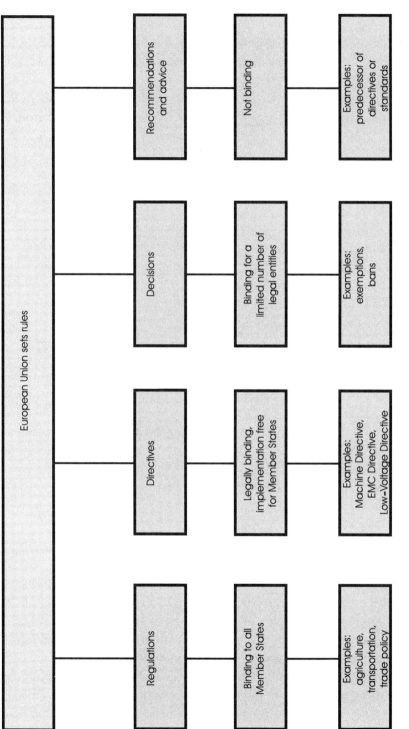

FIGURE 1.3 Hierarchy of Rules and Regulations of the European Union

Regulations are frequently used to formulate the common agricultural policy. Regulations are also used for transport, public procurement, and trade policy.

Regulations are announced in the Official Journal of the European Union and come into effect on the date established in the Regulation.

1.2.2 DIRECTIVES

> A Directive *in the European Union is a European law that is legally binding for every Member State and which is above the laws of the individual Member States. A Directive is not aimed directly at the citizens or companies in that Member State.*

The Member State is free to decide how the Directive must be implemented in the national legislation. However, authority is given to national bodies to select the form and means for realizing the aim of the Directive.

In other words, a Directive has a prevailing power and a uniform effect. This means that Directives are the best means of harmonizing the legislation between the Member States. Within the framework of this manual, this applies in particular to revoking the technical trade restrictions.

1.2.3 Decisions

Decisions are binding in all their parts upon those to whom they are expressly directed. This means that a Decision can apply to both Member States and to private individuals. A Decision can therefore be directed at one legal entity.

Decisions can be both damaging and beneficial. A Decision can have a damaging effect if, for example, a ban is imposed on a national support program (subsidy). But a Decision can also be beneficial, for example, when an exemption is granted from a cartel ban.

The difference between a Regulation and a Decision lies in the fact that a Decision has no general tenor, but relates to a limited number of particularly named or determinable persons. A Decision has binding legal consequences that are directly applicable.

1.2.4 Recommendations and Advice

The European Commission can also provide Recommendations and Advice to the Member States. As expressed by the words themselves, Recommendations and Advice have no binding force. It is therefore up to the Member States to comply, or not to comply, with them. It is possible for instruments of this nature to form the basis for what in future may become a standard or Directive.

1.3 EUROPEAN LEGISLATION FOR PRODUCTS: NEW APPROACH VERSUS OLD APPROACH

Until 1985 the Council of Ministers drew up European Directives in accordance with what is known as the "Old Approach." After 1985 the so-called "New Approach" Directives were drawn up. The reasons for doing this will be explained in the next paragraphs.

1.3.1 Old Approach

This approach meant a Directive was formulated for each product in which all the technical specifications and standards were formulated. The requirements with which a product had to comply were established right down to the nearest detail. However, several shortcomings were involved here:

- Rapid and new technical developments meant that the Directives were often already outdated the moment they came into force. This was because the Directives had been elaborated for every conceivable point and it would therefore take a very long time before the Directive actually materialized.
- A considerable barrier developed during the decision-making procedure for setting up the Directives due to the requirement for unanimity. Under Article 100 of the EEC Treaty it has been established that harmonization Directives must only be accepted with

complete unanimity. Unanimity means that all the members of the Council must give their approval before a decision can be taken.

This involved a lengthy procedure in the Council of Ministers to reach unanimity about the technical details of a product category. It sometimes took years before a definitive EC Directive could be established. Note: The intention of such a Directive was to remove the trade barriers. But this method of approach put in place more trade barriers instead of breaking them down. This led to the European Commission's taking far-reaching initiatives in 1985 to arrive at a more rapid decision-making. The following measures were taken next:

White Paper

First, the White Paper on Completion of the Internal Market was drawn up. This White Paper listed and made an inventory of which juridical steps and measures would be required in order ultimately to realize the EU's objective, namely, the free movement of goods, persons, services, and capital. Furthermore, the White Paper included a time schedule for when the legislation was to come into force. The most recent date quoted for realizing market integration was December 31, 1992. This is why the White Paper's strategy was also propagated as "Europe 1992."

New Approach Directives

To achieve market integration before December 1992, a new method of harmonizing technical Rules and Regulations was essential. In this respect, the Council of Ministers accepted an important proposal by the Commission to switch to what is known as the "New Approach" for harmonizing technical Regulations.

Replacement of the unanimity requirement

Another important development was the acceptance of a new voting procedure for the Council of Ministers. With respect to the legislation for the completion of the internal market it was agreed that decision-making by a weighed, qualified majority would replace the unanimity requirement. The advantage of this was that it speeded up the decision-making process. After all, the individual Member States would no longer be able to use a right of veto.

The most elementary difference between the Old Approach and the New Approach is that under the Directives of the New Approach, a reference is made to standards. Instead of formulating a Directive for each product, Directives are issued for each product group, without a more elaborate formulation of detailed technical specifications.

1.3.2 New Approach

Instead of drawing up European Directives per product, European Directives were drawn up per product group without further explanation of detailed technical specifications. The New Approach Directives impose essential requirements in the field of safety, health, environment, and consumer protection. These essential requirements are formulated in a general way and have been elaborated in the harmonized European standards. For products that have been manufactured in accordance with the harmonized standards, there is an Assumption of Conformity with the Directive.

The European Commission has since authorized and declared as binding a considerable number of Directives. A Directive is authorized when it has been published in the Official Journal of the EU (OJ/EC). This publication also indicates the date when the Directive came into force, i.e., became legally binding.

> *If a product complies with the essential requirements of the applicable New Approach Directives, then the product must be affixed with the CE Marking.*

Unlike the more general meaning of the word "Directive," the observation of the European Directives is therefore not voluntary. The Member States of the European Union are actually obliged to implement the regime of the New Approach Directives in their national legislation. This means that the inhabitants of a Member State have a legal obligation to comply with the applicable Directives. Assume that a manufacturer in the EEA makes a product for his own use, for example, a machine. That

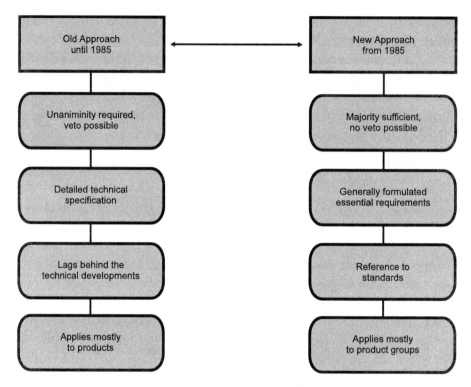

FIGURE 1.4 The "Old Approach" versus the "New Approach" of the EU

machine complies with the Machinery Directive. In such a case, he is also obliged to affix the CE Marking on his machine.

It is also not the case that there is only an obligation to affix the CE Marking in the context of exporting products. Products remaining in the country where they were manufactured in the EEA must also bear the CE Marking.

1.3.3 Advantages and limitations of the New Approach

Figure 1.5 illustrates the advantages and disadvantages of the New Approach. The advantages are:

- *Removal of trade barriers.* One of the greatest advantages of the current approach of the European Union is the realization of generally formulated essential requirements for products and product

groups. These are criteria that apply to all Member States. They are effective in removing the differences and consequently the trade barriers existing between the different countries.
- *Uniform face to the rest of the world.* This also means that Europe presents one uniform face to the rest of the world. The economic superpowers in America and Southeast Asia particularly benefit from straightforward Regulations for trading with Europe.

The disadvantages are:

- *Harmonization goes slowly.* The essential requirements that the European Union sets for products and product groups apply to the fields of safety, health, environment, and consumer protection. These essential requirements are elaborated further by the European standardization organizations like CEN (Comité Européen de Normalisation), ETSI (European Telecommunication Standardization Institute), and CENELEC (Comité Européen de Normalisation Electrotechnique) in their harmonized European standards. However, this process proceeds relatively slowly, sometimes even very sluggishly. National standardization bodies (or, in the case of ETSI, the members) are obliged to cooperate on the drafting of harmonized standards. However, many standardization organizations are confronted by a government engaged in a drive toward privatization, which is therefore tending to withdraw from standardization activities. Today, trade and industry are also concentrating more and more on their core activities, thereby tending to jeopardize their standardization activities.
- *Differences between Member States.* There are often fairly wide differences between the national standards of the various Member States. For example, in the field of electrical safety, the level of safety aimed at in the states of Southern Europe is clearly lower than in Northwest Europe. All this often results in a somewhat mediocre rate of harmonization. This unmistakably conceals one of the European Union's limitations.
- *Existence of grey areas.* Another limitation can be seen in the "grey areas" associated with the European Directives. The Directives always leave some space for interpretation. And where there is possibility for interpretation, the interpretations will be different. This

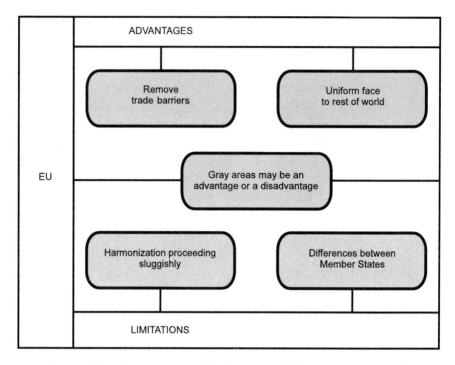

FIGURE 1.5 Advantages and limitations of the New Approach of the European Union

means that differences of opinion arise about what is and what is not the intention of the Directive. Such issues are discussed during the rounds of negotiations between the Member States on the one hand and the day-to-day management of the European Union, the European Commission, on the other. This sometimes results in an explanation being drawn up for a particular Directive, intended to provide entrepreneurs in the EEA with greater clarity. Until such time, it is advisable for individual entrepreneurs in the EEA when doing business in the grey areas to act wherever possible within the spirit of European legislation. If, for example, there is doubt about whether a certain European Directive does or does not apply to a certain product, then it is best to assume that conformity with the Directive will be required. Ensure that you are a step ahead of the Directive; do not hide behind possible ambiguities. If entrepreneurs in the EEA simply sit around and wait for something to happen en masse, they will promote the development and continued existence of the "grey areas."

There is also a favorable aspect to the existence of grey areas: it permits the development of innovatory products. After all, the standards always lag behind the technical developments.

Another limitation of the European Union is the difference in progressiveness between the various Member States. The national governments are of course obliged to implement the European Directives in their national legislation, often within two years. But not all governments succeed in doing this all the time. Of course, the European spirit is more alive in one country than in others. And the observance of the Directives also often differs from one country to the next, and sometimes this even depends on the Directive. As far as the provision of information on European Directives is concerned, the governments of especially the United Kingdom and the Netherlands are particularly active. However, the governments of other Member States are not particularly active in the provision of such information.

But these differences are of little consequence for the entrepreneurs in the EEA. It is important to know that an entrepreneur in the EEA can always fall back on a European Directive, even if the Directive has still not been implemented in the legislation of his particular country.

1.4 STANDARDS AND HARMONIZATION

1.4.1 Definition

In recent years, the unification of Europe has also had far-reaching consequences for the standardization process. Many European standards are drafted in a relatively short period, while the publication of new national standards has almost come to a standstill. Standards do not have the power of legislation. The use of standards is voluntary and not mandatory. Legislation often refers to standards.

The observation of standards is voluntary, unlike the requirements of the Directives, which indeed constitute a legal obligation.

The European Council defines only essential requirements for a product in its Directives. The requirements of a Directive are elaborated further with the aid of standards containing detailed technical specifications.

What is a standard?

> *A standard is a formulated criterion for attaining unity in a field in which diversity is inefficient or unnecessary.*

Standards provide appropriate solutions to certain situations. For example, consider a standard for a screw thread. All the data relating to the screw thread, such as type, diameter, or pitch, are determined in the standard. Products manufactured according to this standard are interchangeable. It is in nobody's interest to depart from this principle, even though, strictly speaking, it is not forbidden.

Standards can be applicable on a national level; an example is the German standards, in which the number is preceded by the letters "DIN." If a standard has been accepted at the European level, the letters "EN" are added.

In addition to the standards mentioned above, there are also international standards such as, for example, the IEC (International Electrotechnical Committee) standards.

The harmonized European standards are of great importance. The national standards of the Member States are replaced by a harmonized European standard.

As concerns the Machinery Directive, more detailed information about the standards for this Directive can be found in Section 2.6 of Chapter 2.

1.4.2 How are standards drafted?

We can define the term "standardization" as follows:

> *Standardization is the process in which rules (standards) are drawn up in order to create order or unity wherever diversity is undesirable and/or unnecessary.*

Standardization takes place with the participation of as many interested parties as possible so that unity that is considered necessary is achieved. An important element is that the groups in society consult each other. In this way, uniform agreements can be made relating to the form, designation, composition, or presentation of products. Standardization therefore has a cost-saving effect since unnecessary diversity of products and production methods is minimized.

Each country within the EEA can use its influence when European standards are being established. The European Commission draws up a mandate to one of the European standardization organizations for the drafting of a standard. A closing date is then established and within that period all interested parties of the EEA can still comment on it. When the standard has been approved, it will be identified by the letters EN. The European standard is then accepted by the national standardization institutes.

In many cases the preliminary design of a standard (before it is approved) will be published in the Official Gazettes and in the specialist journals. Every group in society can react to this preliminary standard by contacting the secretary of the standards committee. The reactions are considered during further consultations, also at European level within the CEN. The national standardization institutions must ensure that the voice of the country's people is always safeguarded within the CEN. The same is true of the CENELEC and ETSI.

The moment that changes are to be implemented in existing technical standards, these are also published. Interested parties can raise objections against these changes. The standardization committee must deal with such an objection.

Table 1.1 shows which institutes are responsible at which level.

1.4.3 European standards and harmonization

The standards and legislation regarding products can differ from one country to the next. It is obvious that this can be an obstruction to free trade within Europe. This is why there is a drive on a European level to what is called technical harmonization.

An EU harmonized standard is a standard that has been accepted by all the national standardization institutes involved. A harmonized standard is announced in the Official Journal of the European Union and is

TABLE 1.1 Overview of standardization organizations and their tasks

	Telecom	Electrical engineering	Other
Global	ITU	IEC IEC-CISPR	ISO
European	ETSI	CENELEC	CEN

ITU	International Telecommunications Union
ETSI	European Telecommunication Standardization Institute
IEC	International Electrotechnical Commission
IEC-CISPR	Comité International Spécial de Perturbations Radioélectrique
CENELEC	Comité Européen de Normalisation Electrotechnique
ISO	International Standardization Organization
CEN	Comité Européen de Normalisation

harmonized from that moment on. The designations of European standards are preceded by letters "EN" (European Norm, i.e., European Standard). Not all European standards are harmonized standards.

The harmonization process is very time-consuming since various national viewpoints have to be supplanted. It is in effect also a political process. This means that harmonized standards usually lag behind the very latest developments in technology. European standards are often still unavailable to the users, while the European Directive referring to them has already become or is about to become a legal obligation.

Everybody is free to ignore the harmonized European standards. The advantage of manufacturing products in compliance with the harmonized European standards, however, is that this will then automatically give rise to an assumption of conformity with the European Directive. The Directive is binding with respect to the result to be achieved for which the standards can offer support. But in themselves, the standards do not constitute any legal obligation or guarantee.

A large number of standards still has to be designed and/or harmonized. To this end, the European Commission has given the task with a mandate to the European standardization institutes—the CEN and the CENELEC. The CEN was founded in 1961 and addresses almost every subject in the field of standardization, with the exception of electrical engineering and telecommunications. The CENELEC is responsible for standardization in the field of electrical engineering and information science. The members of the CEN and the CENELEC originate from the seventeen national standardization institutes of the EEA, the European

Economic Area. Several Technical Committees operate under these institutes, each of which is responsible for a specific area of attention. These Technical Committees are responsible for drawing up the standards.

In 1988, the European Telecommunication Standardization Institute (ETSI) was added to the list of European standardization institutes. ETSI focuses particularly on establishing the standards in the telecommunications sector. The difference between the CEN, CENELEC, and the ETSI is that, besides coming from individual telecommunications organizations in the EEA, ETSI members also come from Turkey, Cyprus, and Malta.

1.5 CONFORMITY-ASSESSMENT PROCEDURES

The CE Marking indicates that the product has been made to comply with the essential requirements set out in the Directive. For a large number of products with a relatively low safety risk, importers themselves may declare that the product meets the essential requirements from the corresponding Directive. This process is called the internal manufacturing inspection.

This situation involves a process of self-certification, and importers meet their responsibility by drawing up the EC declaration of conformity and affixing the CE Marking to the product. Products with a higher risk have to be audited by an external inspection body. These external inspection bodies are also called "Notified Bodies." Only when an importer can have access to the requisite certificate of the external test or inspection organization in the EEA may the CE Marking be affixed.

The European Commission has developed several standard modules for these procedures for interested parties to audit these products for themselves or to have them audited. Directive 93/465/EEC contains the legally binding definition of these procedures, which are known officially as "conformity-assessment procedures."

The gravity of the various types of inspection modules varies depending on the product. Eight modules (modules A through H; see Table 1.2) have been drafted that producers and/or importers can utilize in order eventually to affix the CE Marking on their products. Module A corresponds to the most lenient inspection and module H to the most stringent one. The Commission establishes which module or modules are

TABLE 1.2 Conformance-assessment procedures in Community Legislation. Source: Official Journal of the European Union L 220

	A. (Internal control of production)	B. (Type examination)	C. (Conformity to type)	D. (Production quality assurance)	E. (Product quality assurance)	F. (Product verification)	G. (Unit verification)	H. (Full quality assurance)
DESIGN	Manufacturer - Keeps technical documentation at the disposal of national authorities Aa Intervention of Notified Body	Manufacturer submits to Notified Body - Technical documentation - Type Notified Body - Ascertains conformity with essential requirements - Carries out tests, if necessary - Issues EC type-examination certificate					Manufacturer - Submits technical documentation	EN 29001 Manufacturer - Operates an approved quality system (QS) for design Notified Body - Carries out surveillance of the QS - Verifies conformity of the design[1] - Issues EC design examination certificate[1]
PRODUCTION	A. Manufacturer - Declares conformity with essential requirements - Affixes the CE Marking Aa Notified Body - Tests on specific aspects of the product[1] - Product checks at random intervals[1]		A. Manufacturer - Declares conformity with approved type - Affixes the CE Marking Aa Notified Body - Tests on specific aspects of the product[1] - Product checks at random intervals[1]	EN 29002 Manufacturer - Operates an approved quality system (QS) for production and testing - Declares conformity with approved type - Affixes the CE Marking Notified Body - Approves the QS - Carries out surveillance of the QS	EN 29003 Manufacturer - Operates an approved quality system (QS) for production and testing - Declares conformity with approved type, or to essential requirements - Affixes the CE Marking Notified Body - Approves the QS - Carries out surveillance of the QS	Manufacturer - Declares conformity with approved type, or essential requirements - Affixes the CE Marking Notified Body - Verifies conformity - Issues certificate of conformity	Manufacturer - Submits product - Declares conformity - Affixes the CE Marking The Notified Body - Verifies conformity with essential requirements - Issues certificate of conformity	Manufacturer - Operates an approved QS for production and testing - Declares conformity - Affixes the CE Marking Notified Body - Carries out surveillance of the QS

[1] Supplementary requirements which may be used in specific Directives.

applicable to each Directive. The manufacturer and/or importer usually has a choice of various modules.

1.5.1 Inspection and certification by the importer

Module A (Internal manufacturing inspection)

The first module is intended for products with a low risk. This procedure may be conducted by importers themselves, and this is known as self-certification. Importers themselves draw up the EC declaration of conformity, also known as the manufacturer's declaration. An importer is obliged to keep available the technical documentation of the product for the national authorities for at least ten years after production of the final product. If the manufacturer and/or importer has not manufactured the product according to the harmonized European standards, module A may not be applied.

Module A bis

Manufacturers and/or importers who have not manufactured their products in accordance with the harmonized European standards, but who have applied the essential criteria from the Directive, may make use of this module. The Notified Body within the EEA will then perform tests on specific aspects of the product and conduct random tests at unannounced times. The Notified Body can also have this done on its behalf and under its responsibility by a Designated Laboratory.

1.5.2 Inspection by a Notified Body

For products with an increased risk or which have not been produced according to the standards, a Notified Body must be called in. This applies both to the design phase and the production phase.

Module B (EC type examination)

For products manufactured in quantity this means that four conformity-assessment procedures can be applied. Module B must always be used in combination with module C, D, E, or F, since the CE Marking may never be applied on the basis of module B alone. According to the EC type examination the manufacturer and/or importer must submit a representative type and the technical documentation to the Notified Body. The Notified Body then assesses whether the type is in conformity with the essential requirements from the Directive and, if necessary, conducts tests. Finally, the Notified Body issues a statement relating to the EC type examination.

Module C (Conformity to the type)

Module C entails the following. When module B is complete, the importer draws up an EC declaration of conformity of the approved type. The importer then affixes the CE Marking. Again, for this module, the Notified Body within the EEA tests certain aspects of the product and can conduct random tests at any given moment.

Module D (Production quality assurance)

According to module D, the manufacturer applies an approved quality system for production and inspection, according to standard EN 29002. The Notified Body within the EEA must give its approval for the quality system and monitor this. This module is also used in combination with module B.

Module E (Product quality assurance)

Module E largely resembles module D; the essential difference here is that the focus is not on the quality of the production process, but on the end product, this being within the spirit of standard EN 29003. Here, too, the Notified Body in the EEA has to approve and monitor the quality system. Of course, this module must also be used in conjunction with module B.

Module F (Product verification)

In this module the manufacturer ensures that the production process safeguards that the product complies with the Directives. On the basis of this, the importer draws up the EC declaration of conformity for the approved type. The Notified Body in the EU verifies the EC declaration of conformity and issues a certificate of conformity. Again, module F is also applied after module B.

Module G (Unit verification)

Modules G and H apply to single-unit production. In the design phase the manufacturer will have a Notified Body verify the technical documentation. In the production process the manufacturer submits the product to the Notified Body, and the importer signs the EC declaration of conformity and affixes the CE mark. The Notified Body within the EEA verifies whether each separate product is in conformity with the essential requirements from the Directive and then issues a certificate of conformity.

Module H (Full quality assurance, according to EN 29001)

Much of this module corresponds to module D and module E. However, here the quality system comprises not only the production process and the end product, but also the design process.

As concerns the conformity-assessment procedures for machinery, this will be discussed in Chapter 3.

1.5.3 Verifying observance by the government

Within a certain period (which is specified in the Directive in question), the governments of the Member States in the EEA must transpose the European Directives into national legislation. The national governments within the EEA act as authorized bodies for the verification of the observance of the Directives. If the CE Marking has been wrongfully affixed to the product, the governments within the EEA can take both economic measures and legal procedures. Products, for example, can be taken off the market and large fines can be imposed. Under the Netherlands Economic Offences Act, the Public Prosecutor may even demand a prison sentence for manufacturers or importers who commit such offences.

Audits by the authorized national organizations are carried out by means of random testing or based on accidents or complaints. The function of the market mechanism also plays a role in this. For example, an authorized national organization may (possibly anonymously) receive information from the competitor of an alleged offender. For example, importers who have gone to great lengths to bring their products in conformity with the European Directives and then observe that a competitor has not done so and is therefore at an advantage can report this to the responsible organization in question. In that case, this organization is bound to verify the tip-off by dispatching a detective to the alleged offender.

In future, banks and insurance companies are likely to be playing an increasingly important role as inspection organizations. As money suppliers, the banks will be keeping a close eye on the legal obligations of their customers within the EU, more and more. If a European entrepreneur is insured and an accident occurs after the CE Marking has been affixed unlawfully, the insurance company will take the appropriate steps by not making the payments it would otherwise have been expected to do. After all, the abuse of the CE Marking is always in conflict with the good faith which insurers always take as a basis for their business.

1.6 QUALITY SYSTEMS AND CE MARKING

In the past, taking care of the quality aspects used to focus on the end product apart from the organization that produced the product. Nowadays, quality assessment focuses more on the operation of the organization in its entirety. Focusing on the processes within the organization creates favorable limiting conditions for optimally manufactured products with "zero defects," which are also ready at precisely the right moment, "just in time."

The first systems for quality assessment evolved in the field of the military. From the end of the 1960s, companies wanting to supply products to NATO were obliged to obtain an AQAP (Allied Quality Assurance Publications) certificate. The demand for quality-care systems did not arise in the civilian sector until later. Japan was the pioneer, prompted by American specialists in the field of quality assessment. Europe and America awoke with a start at the success gained by the Japanese with their quality improvements. Entire sectors of industry were actually under threat of falling into Japanese hands; just think of cameras, watches, consumer electronics, and motorbikes.

In 1987, almost twenty years after the AQAP system was set up, five quality-assurance standards were published for civilian use, the now familiar ISO 9000 standards. Companies that have in their possession a quality-care certificate demonstrate with this certificate that they have an efficient organizational form and that they have low failure costs. Failure costs are the costs incurred due to organizational shortcomings within the company. The quality certificate makes no reference, indeed absolutely no reference, to the quality of their products.

> *A company with a quality-care certificate does not by definition supply products of a good quality.*

The certificate is nothing more and nothing less than a recommendation for customers that their order will be processed correctly and on time. The certification according to ISO 9000 is not obligatory.

> *In principle, CE Marking says nothing about the quality of the product.*

The CE Marking indicates that the product complies with the essential requirements relating to the safety, health, environment, and consumer protection of the user.

The following two examples illustrate these statements:

1. An electrotechnical installation company is certificated according to ISO 9001. This certificate does not mean that the company will also carry out good installation work. During the certification process, the focus is only on whether the company has the appropriate resources to design and install electrotechnical systems, whether the personnel are well equipped and trained, and whether the correct accounting procedures are observed. The actual installation work does not form part of the certification procedure, since the result of the installation depends each time on the individual approach of the technicians in question.

2. A child's bucket with holes in the bottom may well bear the CE Marking according to the Directive governing children's toys. This

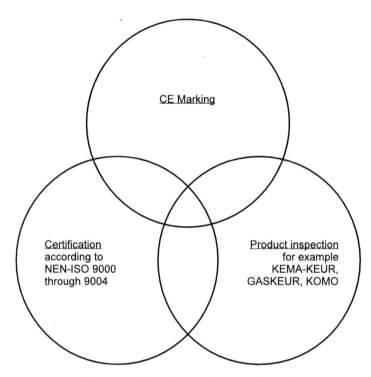

FIGURE 1.6 CE Marking, product inspection, and quality-care certification complement each other and partly overlap each other

demonstrates that the bucket conforms to the essential requirements in the field of safety, health, environment, and consumer protection. But the bucket's quality in respect of its ability to transport sand or water is clearly appalling.

If we wish to provide our customers with assurance about the functional quality of a product, we will need to obtain a voluntary inspection mark; for example, in the Netherlands these include KEMA-KEUR, Gaskeur, KOMO, GS. These inspection marks provide guarantees that products conform to safety and functional requirements on the long term. For products with an inspection mark, it is sometimes also the case that they meet the essential requirements according to applicable European Directives. In such a case, on the basis of this, little extra effort is required to affix the CE Marking in accordance with the Directive. But it is also sometimes the case with some inspection marks that they do not focus at all on the essential requirements according to other Directives.

In the case of an electrotechnical safety inspection, for example, no account is taken of the EMC aspects of the product. It is advisable to approach the relevant inspection organization where individual cases are concerned.

And what is the relation between ISO 9000 and CE Marking? A certificated company is not automatically authorized to affix the CE Marking. Although the company will find its quality-care certificate convenient en route to the CE Marking, there is no further causal connection with the CE Marking. The demonstrable quality of the organization in the case of products made in quantity will at least help to maintain the conformity of the products with the essential requirements in the EC Directive(s). Indeed, in the case of quantity or mass production, there must be an analysis of which deviation may arise in the product and which tolerances are permitted in order to continue to comply with the essential requirements of the relevant Directive. If certificated companies develop and produce products that comply with the essential requirements of the Directive in question, this will also apply to quantity production.

1.7 CONSEQUENCES FOR THE AMERICAN MANUFACTURER

1.7.1 Introduction

New requirements now apply to industrial and consumer products in all European countries. These requirements relate to health and safety, the environment, and consumer protection and have been drawn up by the European Council in various Directives applying to more than 40% of industrial products. The compliance of a product with the relevant Directive is indicated by affixing the CE Marking. Several Directives may apply to one product. A product with the CE Marking can be placed on the market without restriction in all the countries of the EEA. The advantage of this is that one uniform legislation now applies to all European Member States.

If American companies want to sell their products on the European market, they are obliged to conform with these European Directives. Products not complying with applicable Directives may not be placed on the EEA market. A Directive has a specific area of application, or scope, and separate dates on which it is introduced and becomes mandatory.

This is why it is very important for American companies to be aware and stay aware of the contents and the consequences of the various Directives.

The central element of the European Directives is user safety. If damage or injury is sustained from a particular product, the user may hold the supplier liable. This is why the Directives must be observed correctly. A safe product that complies with the essential requirements of the corresponding Directives minimizes the risk of a claim for damages.

This chapter discusses, in general, where the responsibilities lie and what steps American manufacturers must take to carry out the certification procedures correctly. As concerns the detailed formalities for the Machinery Directive, we refer to Chapter 3.

1.7.2 Responsibilities

Before the CE Marking may be affixed to a product, several requirements and formalities must be met. Manufacturing the product in conformity with the European Directive is the responsibility of the American producer. The compilation of the Technical File (TF; or Technical Construction File, TCF) plays an important role here. The Technical Construction File largely contains inspection reports and investigation results which show that the product characteristics fall within the prescribed limits. American producers must have this file at their disposal to demonstrate compliance with the Directives. Contractual agreements will then be reached with the European importer in respect of the Technical Construction File and liability. These agreements are necessary because the legislation holds the European importer responsible for the CE Marking.

If an accident occurs in the EEA due to a product originating in America, a European legal entity will be held responsible for this. An EC declaration of conformity must be provided for every product containing the name and address of the European importer. The EC Declaration of Conformity is a document in which the manufacturer in the EEA, the manufacturer's authorized agent domiciled in the EEA, or the importer officially declares that the product complies with all the essential requirements of the Directives currently mandatory. This makes it possible to trace the company responsible in the EEA.

It will be necessary to reach contractual agreement between the American company and the authorized agent or importer established in the EEA in order to cover a liability issue, for example, how a situation will be handled in the event of a claim for damages as the result of an accident.

1.7.3 Steps for exporting from the USA to the EEA

Before the CE Marking may be affixed to a product, the product must comply with the essential requirements of the Directive(s), and compliance is required with the formalities indicated by the Directive(s). Figure 1.7 outlines the steps that a producer in the USA must follow before exporting a product to the EEA.

Obligations for the producer in the USA

1. The American producer must ensure that the technical aspects of a product comply with the essential requirements of the Directive(s). The requirements from the applicable Directives must first be integrated into the design. For quantity manufacture, one must implement internal measures to ensure that the product remains in conformity with the provisions of the Directive.

2. A Technical File must be drawn up containing all the information required for demonstrating conformity of the product and the production process. This information includes, for example, test reports, diagrams, calculations, and specifications. The American producer must also provide all the operating data that has to be supplied with the product. This data includes, for example, the user manual, the installation instructions, the safety instructions, and maintenance manuals.

3. If the Directive prescribes that products have to be tested in accordance with certain Directive(s), this can be done at the European Notified Bodies or at a testing institute in the USA having a subcontracting agreement with a European Notified Body. In the last case, the assessment of the test result will be done by the corresponding Notified or Competent Body in Europe (see step 5).

Obligations for the importer in Europe

4. The European importer must verify conformity with the Directive(s) and take formal responsibility for the CE Marking. This means the importer must be able to justify compliance with the Directives. An EC

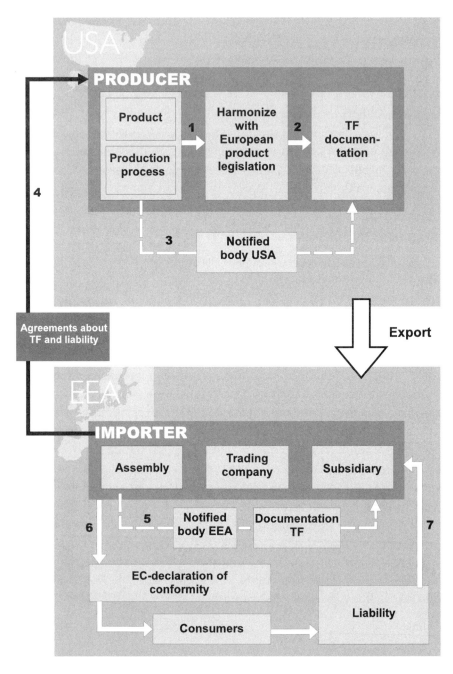

FIGURE: 1.7 Steps for exporting from the USA to the EEA

declaration of conformity must be drawn up, in which the European importer officially declares the compliance with the Directive(s). The European importer must therefore be able to inspect the Technical File. Contractual agreements must be reached regarding the way in which the Technical File is made available. The sales contract must also contain articles relating to liability.

5. The European importer fulfills the final formalities in the procedure for the CE Marking. If products have been tested in the USA, then official approval will be issued by a Notified Body in the EEA. All user information must be presented in the national language of the country to which the product is to be imported.

6. The European importer draws up the EC declaration of conformity and signs it as the responsible company. The product is supplied with the CE Marking, the EC declaration of conformity, and all user information in the national language of the relevant European country.

7. Formal handling of liability issues is carried out in Europe.

1.7.4 Mutual Recognition Agreements (MRAs)

The governments of the United States and the European Union have signed a comprehensive agreement of mutual recognition (MRA). This agreement allows an exporter to test and certify his or her product on native soil and have that certification recognized in the territory of an importer's country. These agreements establish a way to trust in the competence of one another's conformity assessment systems.

MRAs save US exporters to the European Union a great deal of money and time, break down technical obstacles, and ease trade in both directions. Under an MRA, a manufacturer can meet both US and EU standards by undergoing testing, inspection, or certification procedures in whichever country is most convenient.

For example, a certifying body in the United States may be deemed competent to test to European standards and certify that products conform to European laws and regulations. The certificate of that body located in the United States will be accepted by European authorities. Likewise, there will be bodies in the European Union that will be deemed competent to test and certify to US standards and laws. This means that if your product falls within the range of the agreements, you are able to test and certify in the United States and ship your product to the European Union without further assessment, and vice versa. The saving will be enormous.

The products involved are those products which require third-party certification, either within the United States or within the European Union, or in both. The product areas covered are:

- Telecommunications equipment
- All products that must conform with electrical safety laws and electromagnetic compatibility laws
- Pharmaceuticals
- Medical devices
- Recreational craft

1.8 SUMMARY OF NEW APPROACH DIRECTIVES AND PROPOSALS

1.8.1 Overview of all New Approach Directives

Table 1.3 shows an overview of the starting dates and transition dates of all the New Approach Directives adopted to date.

1.8.1.1 Active implantable medical devices—90/385/EEC.

The active implantable medical devices Directive sets essential requirements with respect to medical devices powered by an electrical energy source. These medical devices are produced for implantation in the human body. The active implantable medical devices Directive came into force on 1 January 1993, and it has been a legal obligation since 1 January 1995. Now, active implants may no longer be traded without the CE Marking.

The essential requirement in the Directive can be divided into two categories: the general requirements and requirements that apply to the design and construction of the active implant. These latter requirements focus on the technical safety, biocompatibility, and sterility of the active implant.

Examples of active implants are, among other things, implantable neurostimulators, hearing aids, and insulin pumps. However, the Directive defines the term "active implant" quite broadly, and not only the implant

TABLE 1.3 Overview of all New Approach Directives

Directive	Starting date	Transition period until
Active implantable medical devices	01-01-1993	01-01-1995
Equipment and protective systems intended for use in a potentially explosive environment	01-01-1996	01-01-2003
Simple pressure vessels	07-01-1990	07-01-1992
Electromagnetic compatibility Directive	01-01-1992	01-01-1996
Appliances burning gaseous fuels	01-01-1992	01-01-1996
Low voltage Directive	01-01-1995	01-01-1997
Machinery Directive	01-01-1993	01-01-1995
Medical devices	01-01-1995	06-14-1998
Non-automatic weighing machines	01-01-1993	01-01-2003
Personal protective equipment	07-01-1992	07-01-1995
Telecommunications terminal equipment and satellite earth station equipment	01-04-1998	Via CTRs[1]
Efficiency requirements for new hot-water boilers fired with liquid or gaseous fuels	01-01-1994	01-01-1998
Safety of toys	01-01-1990	No transition period
Domestic electric refrigerators, deep freezers	10-08-1996	09-03-1999
Building materials	06-27-1991	No transition period
Pleasure Craft	06-16-1996	06-16-1998
Explosives for Civil Use	01-01-1995	01-01-2003
Pressure Equipment	11-29-1999	05-29-2002
Lifts	07-01-1997	07-01-1999
High-speed trains	10-08-1996	No transition period

[1]This Directive is the only Directive which uses "Common Technical Requirements" (CTRs).

itself must bear the CE Marking, but also the associated accessories and software without which the implant cannot be operated.

1.8.1.2 Equipment and protective systems intended for use in a potentially explosive environment—94/009/EEC. The Directive equipment and protective systems intended for use in a potentially explosive environment sets essential safety and health requirements. These requirements relate to the design and construction of equipment and security systems intended for use in locations that may have a risk of explosion. The scope of this Directive also includes the

safety, inspection, and regulatory provisions intended for use outside locations where there may be a risk of explosion, but which are a prerequisite for or contribute to the safe operation of equipment and security systems relating to explosion hazards.

The Directive came into force on 1 March 1996 and will become a legal obligation on 1 July 2003.

1.8.1.3 Simple pressure vessels—87/404/EEC and 90/488/EEC.

The simple pressure vessels Directive sets essential safety and health requirements for quantity-produced simple pressure vessels. The safety requirements formulated in the Directive are applicable to the use of materials, the design, and the manufacturing method for simple pressure vessels and their use.

The simple pressure vessels Directive came into force on 1 July 1990 and was subject to a transition period until 1 July 1992. The Directive defines a pressure vessel as follows: "a series [quantity]-produced welded pressure vessel with an internal overpressure of more than 0.5 bar, intended to contain either air or nitrogen." A simple pressure vessel relates therefore to pressure vessels intended for air or nitrogen. These are usually welded pressure vessels. These pressure vessels are not exposed to fire or heating.

Exceptions to these simple pressure vessels are formed by:

- Pressure vessels intended for use in nuclear installations in which faults may lead to the release of radioactivity
- Fire extinguishers
- Pressure vessels intended for the installation on or for the propulsion of ships and aircraft

1.8.1.4 Electromagnetic Compatibility—(EMC) 89/336/EEC and 92/31/EEC.

The Electromagnetic Compatibility Directive sets the essential requirements for electrical and electronic equipment that may disturb or even be disturbed by other equipment. This Directive relates to all electrical or electronic equipment.

The EMC Directive came into force on 1 January 1992, and it became a legal obligation on 1 January 1996. This means that from that moment on, no more equipment may be placed on the market or put into operation without the CE Marking. The CE Marking may be omitted from equipment that was already placed on the market before that date. It will very rarely be the case that only the EMC Directive is applicable to a product. With electrically driven machines, for example, a combination of the Machinery Directive, the Low-Voltage Directive, and the EMC Directive will often apply.

1.8.1.5 Appliances burning gaseous fuels—90/396/EEC.

The appliances burning gaseous fuels Directive formulates essential requirements for gas-fired appliances and their components as well as for fan-assisted combustion units and the heating boilers for these. There are essential general conditions and requirements relating to the materials, design, and construction. The appliances burning gaseous fuels Directive came into force on 1 January 1992 and became a legal obligation on 1 January 1996. From that moment on, no more appliances burning gaseous fuels may be placed on the market or installed without the CE Marking.

The scope of the Directive concerns equipment intended for:

- Cooking, heating, hot-water production, cooling, lighting, or washing. Here, the maximum water temperature may not exceed 1050°C.
- Fan-assisted combustion units and heat generators equipped with this kind of combustion unit.
- Security, inspection, and regulation equipment and components marketed separately for professional use which are intended for installation in a gas appliance or to be assembled to form a gas appliance.

Appliances burning gaseous fuels intended specifically for industrial processes are not covered by this Directive.

1.8.1.6 Low Voltage—73/23/EEC.

The Low-Voltage Directive sets essential requirements with respect to the electrical safety of electrotechnical products. The Low-Voltage Directive was established as early as 1973. This makes it the first Directive formulated according to the method of the "New Approach." This means only the general requirements are drawn up in the Directive, without the technical details being formulated specifically. However, the Low-Voltage Directive of 1973 does not include the obligation to affix the CE Marking. The Directive was changed to this end in 1993. As of 1 January 1997, this obliges every manufacturer or importer to affix the CE Marking to the products covered by the scope of the Low-Voltage Directive.

The Low-Voltage Directive is applicable to all equipment or machines requiring a voltage that lies in the range:

50–1000 V for AC voltage, or
75–1500 V for DC voltage

An important element of this Directive is danger of electric shocks with all the consequences it entails. Besides the electric-shock hazard, distinction is also made between other dangers. Hazardous situations may result from not being able to withstand deviations that can (may) occur on the power network.

1.8.1.7 Machinery—89/392/EEC, 91/386/EEC, and 93/44/EEC.

The machinery Directive sets essential safety and health requirements relating to the design and construction of machines and safety components. The Machinery Directive came into force on 1 January 1993 and became a legal obligation on 1 January 1995. This means that from that date, no machinery may be placed on the market or installed if it does not bear the CE Marking.

The Machinery Directive assigns a very broad description to the concept of a machine. The scope of the Machinery Directive therefore ranges from a simple jack used to lift a car to a complicated packaging line. The essential requirements formulated in the Directive relate to the safety of the product. Here, the CE Marking indicates that the product has been made to conform with the essential safety requirements set in the Directive.

Please note that Machinery Directive 89/392/EEC and amendments 91/386/EEC (only Article 1), 93/44/EEC, and 93/68/EEC (only Article 6) were repealed in August 1998 and replaced by Directive 98/37/EC. This Directive integrates the previous Directive 89/392/EEC and its amendments and has been incorporated in Appendix V of this book.

1.8.1.8 Medical devices—93/42/EEC.
The medical devices Directive sets general requirements and requirements relating to the design and construction of medical devices. The Directive applies to all medical devices, except for active implantable medical devices. The separate active implantable medical devices Directive is applicable to them. The Directive defines the term "medical device" as follows: appliance for alleviating the effects of a handicap or illness.

An important point of departure is that the focus is on the primary function of a product within the framework of the Directive. For example, spectacles fall under the scope of the Directive since their primary function is aimed at counteracting partial sightedness. The primary function of a pair of sunglasses is to protect against UV radiation and may have a secondary function assisting partial sightedness. Consequently, a pair of sunglasses is not a medical appliance as described by the Directive.

The medical devices Directive came into force on 1 January 1995 with a transition period until 14 June 1998. An exception has been made for clinical thermometers, which have a transition period until 30 June 2004.

1.8.1.9 Non-automatic weighing machines-90/384/EEC.
The non-automatic weighing machines Directive sets out essential requirements with which non-automatic weighing machines must comply. "Non-automatic weighing machines" refers to those weighing instruments in which the intervention of an operator is required for the weighing activity. The fundamental requirements can be distinguished by metrological requirements and requirements relating to the design and construction of non-automatic weighing machines, but also, for example, EMC requirements.

The Directive came into force on 1 January 1993 and will become a legal obligation on 1 January, 2003.

1.8.1.10 Personal protective equipment—89/686/EEC.
The personal protective equipment Directive sets essential requirements for pieces of equipment or items intended to be worn by persons or carried by them to protect against hazards that are a danger to health and/or safety. The essential requirements are composed of general requirements applicable to all protective equipment and which relate to the principles of design, innocuity of the protective equipment, comfort, and appropri-

ateness and the manufacturer's instructions for use. Protective equipment may be unattached products, but can also form part of a greater entity.

The following fall outside the scope of the Directive:

- Protective equipment for military applications
- Protective equipment for self-defence
- Protective equipment for use by consumers to protect against moisture, water, heat, certain weather conditions
- Protective equipment intended for rescue activities on board ships or aircraft

The personal protective equipment Directive came into force on 1 July 1992 and became a legal obligation on 1 July 1995.

1.8.1.11 Telecommunications terminal equipment and satellite earth station equipment—98/13/EC.
The text of the telecommunications terminal equipment Directive 91/263/EEC and the satellite earth station equipment Directive 93/97/EEC were consolidated into one Directive, 98/13/EC. This Directive came into force 1 April 1998 and sets essential requirements for all terminal equipment that can be connected to the public telecommunications network and satellite earth station equipment. This Directive has no transition period; the transition periods have been included in the CTRs (Common Technical Requirements). CTRs are the equivalent of harmonized standards for telecom equipment. All equipment which is within the scope of this Directive may also fall within the scope of the EMC and Low-Voltage Directive. A proposal is currently before the European Commission to issue a new Directive which will replace the 98/13/EC.

1.8.1.12 Efficiency requirements for new hot-water boilers fired with liquid or gaseous fuels—92/42/EEC.
The efficiency requirements for new hot-water boilers fired with liquid or gaseous fuels Directive sets no essential requirements as is the case with the other Directives. However, specific requirements are set for the efficiency for new hot-water boilers fired with liquid or gaseous fuels.

The efficiency requirements for new hot-water boilers fired with liquid or gaseous fuels Directive has been in force since 1 January 1994 and had a transition period until 1 January 1998. The specific requirements in the Directive relate to the type of appliance, i.e., the standard boiler, the low-temperature boiler, and the condensing boiler. Every central-heating boiler, gas and/or oil-fired, with a capacity between 4 and 400 kW falls under the scope of the Directive. This also applies to unattached, fan-assisted combustion units and the boilers fitted with them. The European standards establish how the efficiencies can be determined.

1.8.1.13 Safety of toys—88/378/EEC.
The safety of toys Directive sets fundamental health and safety requirements relating to playthings. The Directive relates to, among other things, toys, which means any product that has been designed or intended for use by children younger than fourteen years of age when playing.

The toys Directive already came into force on 1 January 1990 and has no transition period. This means that no toy may be placed on the market and put to use without the CE Marking.

The essential safety requirements for toys comprise general requirements relating to the design, the construction, or the composition of the toys. Further, there are essential safety requirements relating to special hazards, such as the physical and mechanical properties, flammability, chemical properties, electrical properties, hygiene, and radioactivity.

1.8.1.14 Domestic electric refrigerators, deep freezers—96/57/EC.
The Directive domestic electric refrigerators and deep freezers concerns standards which are formulated for the energy efficiency of domestic electrical refrigerators, deep freezers, and combinations of these. Domestic refrigerators, storage spaces for deep-frozen foodstuffs, deep freezers, and combinations of these connected to the electrical power supply network fall under the scope of the Directive. The Directive describes procedures for calculating the maximum permitted electricity consumption for a certain type of cooling equipment. Furthermore, the inspection of the conformity is also described. The Directive came into force 8 October 1996 and has a transition period until 3 September 1999.

1.8.1.15 Building materials—89/106/EEC. The building materials Directive contains essential requirements relating to "materials intended for construction." "Materials intended for construction" are defined in the Directive as those products that are manufactured to form a permanent part of structures. Structures refers to both buildings and works of art.

The building materials Directive came into force on 27 June 1991. At present, the building materials Directive is in a transition period since the European Commission still has to establish when the Directive will become a legal obligation.

Materials intended for construction must comply with the fundamental regulations during an economically relevant lifetime and provided that regular maintenance is carried out. These fundamental regulations are subdivided into six aspects, namely:

- Mechanical strength and stability
- Fire safety
- Hygiene, health, and environment
- Safety of use
- Sound nuisance
- Energy savings and heat retention

1.8.1.16 Pleasure craft—94/25/EEC. The pleasure craft Directive sets essential ressential requirements for the design and construction of pleasure crafts. The Directive defines a pleasure boat as "any craft intended for sport and leisure purposes, regardless of the type or the means of propulsion, with a hull length of 2.5 to 24 meters, measured according to the appropriate harmonized standards." The area of application of the Directive applies to pleasure craft, partially completed pleasure craft and loose and assembled components.

The pleasure craft Directive came into force on 16 June 1996 and will have a transition period until 16 June 1998. After that date, manufacturers and importers may no longer place pleasure craft on the market that do not comply with the essential requirements of the Directive.

Several Directives may also apply to a pleasure boat, for example the EMC, telecommunications terminal equipment, personal protective equipment, gas appliances, and simple pressure vessels Directives. If the aforementioned Directives have already become a legal obligation, the manufacturer or importer must also take into account the essential requirements of these Directives (where applicable). So it is always necessary to verify exactly which Directives are applicable here.

1.8.1.17 Explosives for civil use—93/15/EEC.
The explosives for civil use Directive sets essential safety requirements for explosives. The United Nations Recommendations on the Transport of Dangerous Goods describe all substances and materials that may be considered to be "explosives." Explosives (including munitions) that are intended for use in accordance with national legislation by the armed forces or the police therefore do not fall within the scope of application of the Directive.

The explosives for civil use Directive came into force on 1 January 1995 and will become a legal obligation on 1 January 2003. So at present the Directive is in a transition period, although manufacturers and importers of explosives for civil use are already able to bring their products in line with the essential safety requirements of the Directive.

The essential safety requirements for explosives for civil use are divided into general requirements and the special requirements. The general requirements relate to, among other things, the design, so that the explosives constitute only a minimal danger for the health and safety of persons. The special requirements relate to, for example, the shock and friction sensitivity and the chemical purity of the explosive.

1.8.1.18 Pressure equipment—97/23/EC.
The Directive for pressure equipment sets essential requirements for the design and manufacture of "equipment subjected to high pressure" or "pressure equipment" that is subjected to a permissible PS overpressure greater than 0.5 bar or less than –0.5 bar. The scope of application of the proposal for pressure equipment covers the following:

- Pressure vessels
- Pipelines
- Assemblies
- Accessories

The Directive pressure equipment will come into force 29 November 1999 and will have a transition period until 29 May 2002.

1.8.1.19 Lifts—95/16/EC. The Lifts Directive sets essential health and safety requirements for the design and construction of elevators and safety components. The Lifts Directive applies to permanently installed elevators in buildings and structures and relates specifically to personnel elevators and elevators that are used to carry personnel and goods. Not included in the Directive are cable installations, elevators for military purposes, mine elevators, theatre hoists, elevators that are installed in means of transport, elevators that are connected to a machine, rack railways, and construction elevators.

Lifts Directive 95/16/EC took effect on 1 July 1997 and had a transition period until 30 June 1999.

1.8.1.20 High-speed trains—96/48/EC. The high-speed trains Directive sets requirements relating to the interoperability of the European network for high-speed trains. The objective of the Directive is to establish conditions that must be met in order to realize such interoperability. The Directive concerns the design, development, gradual implementation, and operation of the high-speed-train network. The high-speed trains Directive sets essential requirements with which the European network or each part of it, the subsystems and the constituent parts, must comply.

The Directive came into force on 8 October 1996 and has no transition period.

1.8.1.21 Measuring equipment—proposal. At the end of 1995, the proposal for the measuring equipment Directive was still in preparation by the European Commission. It will subsequently be presented to the Council of Ministers. The measuring equipment proposal cannot be published until the Council has approved it.

1.8.1.22 Cableway installations for personnel transport—proposal. This relates to a proposal for a Directive relating to cableway installations for personnel transport. The proposal sets essential requirements.

At present, cableway installations for personnel transport in the Member States are subject to legal regulations relating to their operation and their safety when placed into operation as well as aspects relating to environmental protection and environmental planning. The proposal for a Directive for cable installation for personnel transport divides the installations in question into five groups, namely funicular railways, cable cars, cabin lifts, chair lifts, and ski drag lifts. The proposal for the Directive sets essential requirements relating to the system in its entirety, in which account should be taken of the end result of the assembly of the components.

The New Approach cable installation Directive for personnel transport has not yet been discussed by the Council.

1.8.1.23 Precious metals—proposal.
This relates to a proposal for a Directive for working with precious metals. The proposal sets essential requirements relating to the content for working with precious metals for consumers. Content refers to the fine precious metal content, expressed in thousandths of the total mass of the alloy in question. An alloy is composed of a fixed solution of precious metals and one or more other metals. The items of work containing precious metals must therefore bear a hallmark.

The scope of applications covers precious metals such as pure platinum, gold, palladium, silver, or alloys of these metals, and also mixed work comprising components of precious metals and components of base metal. Exceptions from the proposal for the Directive include work items of precious metal for dentures and medical use, and musical instruments or parts thereof manufactured from precious metals.

The proposal for a Directive relating to works of precious metals was submitted by the Commission on 14 October 1993.

1.8.1.24 In vitro diagnostics—proposal.
This relates to a proposal for a Directive relating to medical appliances for in vitro diagnostics and their accessories. The proposal refers to the accessories as fully-fledged medical appliances for in vitro diagnostics.

The proposal describes a medical appliance for in vitro diagnostics as follows: "each medical appliance that is a reagent, a reactive product, a calibration material, a control material, a kit, an instrument, a device, an appliance or a system that is used separately or in combination and which the manufacturer intends to be used for the in vitro investigation of specimens originating from the human body, including donor blood and tissue, exclusively or principally with the objective of providing information on the physiological condition, the health, the illness or a congenital defect or for determining the safety thereof and the level of compatibility with potential receptors."

The proposal for a Directive relating to medical appliances for in vitro diagnosis was submitted by the Commission on 19 April 1995. The draft is still being processed.

1.8.2 The CE Marking Directive

CE Marking Directive 93/68/EEC came into force in the EU on 22 July 1993. At that time there were already twelve Directives within the framework of the New Approach. Detailed changes were found to be necessary in the already established Directives. These changes related mainly to the affixing and use of the CE Marking.

The CE Marking Directive is the document that officially uses the term "CE Marking" for the first time. Until then, the term EC-symbol was used. It also strictly defines the graphic design of the letters "CE."

With the introduction of the CE Marking Directive, the affixing of the CE Marking to products harmonized with the corresponding Directive has become a legal obligation.

The twelve Directives changed by the CE Marking Directive are presented below:

- 87/404/EEC: Simple pressure vessels
- 88/378/EEC: Safety of toys
- 89/106/EEC: Construction products
- 89/336/EEC: Electromagnetic compatibility
- 89/392/EEC: Machinery
- 89/686/EEC: Personal protective equipment
- 90/384/EEC: Non-automatic weighing machines
- 90/385/EEC: Active implantable medical devices
- 90/396/EEC: Appliances burning gaseous fuels
- 91/263/EEC: Telecommunications terminal equipment
- 92/42/EEC: Efficiency requirements for new hot-water boilers fired with liquid or gaseous fuels

- 73/23/EEC: Electrical equipment intended for use within certain voltage limits

Council Decision number 93/465/EEC was formulated simultaneously with the CE Marking Directive. This Decision was adopted within the EU on 22 July 1993 and contains conditions for the application of the CE Marking and the harmonization procedures applied in the New Approach.

2
Machinery Directive—General

2.1 THE DIRECTIVES (89/392/EEC, 91/386/EEC, 93/44/EEC, AND 93/68/EEC)

The Machinery Directive contains fundamental health and safety requirements relating to the design and construction of machines and safety components. The Machinery Directive came into force effective 1 January 1993 and became a legal requirement effective 1 January 1995. From this last date, machines may no longer be sold or commissioned within the EEA unless they have a CE Marking.

The Machinery Directive was amended for the first time in 1991. As a result of this amendment, the following also fall under the Machinery Directive:

- Machines for lifting or moving persons
- Safety components
- Falling object protective structure (FOPS) and rollover protective structure (ROPS)

The transition period for this amendment ended on 1 January 1997. The Machinery Directive has been amended two more times since, by Directive 93/44/EEC and Directive 93/68/EEC.

Please note that Machinery Directive 89/392/EEC and amendments 91/386/EEC (only Article 1), 93/44/EEC, and 93/68/EEC (only Article 6) were repealed in August 1998 and replaced by Directive 98/37/EC. This Directive integrates the previous Directive 89/392/EEC and its amendments and has been incorporated in Appendix V of this book.

2.2 AREA OF APPLICATION

The Machinery Directive applies to:

1. Machines as described in the Machinery Directive, Article I section 2:

An assembly of linked parts or components, at least one of which moves, with the appropriate actuators, control and power circuits, etc., joined together for a specific application, in particular for the processing, treatment, moving or packaging of a material

An assembly of machines which, in order to achieve the same end, are arranged and controlled so that they function as an integral whole

[A piece of] interchangeable equipment modifying the function of a machine, which is placed on the market for the purpose of being assembled with a machine or a series of different machines or with a tractor by the operator himself in so far as this equipment is not a spare part or a tool

2. *Safety components*, if they are placed on the market separately, are also subject to the Machinery Directive and are defined as follows:

A component, provided that it is not interchangeable equipment, which the manufacturer or his authorized representative established in the Community places on the market to fulfil a safety function when in use and the failure or malfunctioning of which endangers the safety or health of exposed persons

It is also important to realize that the Machinery Directive does not concern itself with the correct operation or performance of the machine, provided safety is not at issue as a result.

In order to clarify the above, the terms "interchangeable equipment," "safety component," "composition of machines," and "machine parts" will be explained in more detail.

Interchangeable equipment

A piece of interchangeable equipment is a tool or component with which the original function of the machine can be changed. The feature of such equipment is that it can be connected and disconnected by the user.

The interchangeable equipment should be certified as if it were a machine.

In a technical sense it is a machine, albeit that the drive is supplied by an external power source. The prerequisite is that the equipment is actually interchangeable. In other words, the installation operations by the user should be minimal and uncomplicated so that no legal obligations can be placed on the user.

An important criterion for interchangeable equipment can therefore be couched in terms of compatibility: the equipment should naturally be suitable for connecting to a basic machine.

In the consumer sphere, the example of a drill is obvious. This machine is a tool for drilling, but it is also a basic drive for a large number of interchangeable tools such as a pump, a sander, or a saw.

Safety component

Safety components may be essential for the safe operation of a machine. They are used by manufacturers of machines which are in principle unsafe in order to make the machine comply with the requirements of the Directive and to increase the safety level of the machines. Some examples of safety components are a fixed or mobile protection or two-button operation (for example, when supplied on a hydraulic press).

A safety component may not be given the CE Marking.

Composition of machines

"Composition of machines" means an installation or production line which consists of two or more machines. This means the application or setting up and preparation for use of a machine or part. Examples of installations are robotic and automated workstations.

Machine parts and semi-finished products

A product is a machine part or a semi-finished product if it meets the following requirements:

- It is not intended for one specific application.
- It can be used as part of a larger whole with a specific function.

- It cannot be used operating on its own.

Examples are electric motors, combustion engines, transmissions, and incomplete machines.

A manufacturer's declaration should be supplied with a machine part under Annex II, Point B, of the Machinery Directive. A manufacturer of a machine part is not obliged to supply such a declaration, but it is advisable. With the manufacturer's declaration, the producer is imposing a prohibition on commissioning of the machine part. This means that the part must not be used before it has been built into or onto a whole which complies with the Machinery Directive.

2.3 EXCEPTIONS

Exceptions have been made in the sphere of operation of the Directive. These are indeed machines in the sense of the Directive, but these machines do not fall under the scope of the Directive. The following do not fall under the Directive:

- Machines which are excepted in the Directive (Article 1, Section 3)
- Machines for which the risks which are specified in this Directive fall under the sphere of operation of a particular Directive (Article 1, Section 4)
- Machines for which the risks are mainly of electrical origin (Article 1, Section 5)
- Machines with which no major risks are associated

This last exception is not stated explicitly in the Directive, but follows from the intent of the Directive. The safety requirements from Annex I of the Machinery Directive should only be applied if the risk analysis shows that the risks to which the safety requirements are geared arise.

The Machinery Directive also contains an entire list of exceptions in Article 1, Section 3. The following machines appear in this list:

- Machinery whose only power source is directly applied manual effort unless it is a machine used for lifting or lowering loads
- Machinery for medical use used in direct contact with patients

- Special equipment for use in fairgrounds and/or amusement parks
- Steam boilers, tanks, and pressure vessels
- Machinery specially designed, or put into service, for nuclear purposes which, in the event of failure, results in an emission of radioactivity
- Radioactive sources forming part of a machine
- Firearms
- Storage tanks and pipelines for gasoline, diesel fuel, inflammable liquids, and dangerous substances
- Means of transport, i.e., vehicles and their trailers intended solely for transporting passengers by air or on road, rail, or water networks, as well as means of transport in so far as such means are designed for transporting goods by air, on public road or rail networks, or on water; vehicles used in the mineral extraction industry shall not be excluded
- Seagoing vessels and mobile offshore units together with equipment on board such vessels or units
- Cableways, including funicular railways, for the public or private transportation of persons
- Agricultural and forestry tractors, as defined in Article 1, Section 1, of Council Directive 74/150/EEC of 4 March 1974 on the approximation of the laws of the Member States relating to the type-approval of wheeled agricultural or forestry tractors, as last amended by Directive 88/297/EEC
- Machines specially designed and constructed for military or police purposes
- Lifts which permanently serve specific levels of buildings and constructions, having a car moving between guides which are rigid and inclined at an angle of more than 15 degrees to the horizontal and designed for the transport of:
    - Persons
    - Persons and goods
    - Goods alone if the car is accessible, which is to say, a person may enter it without difficulty and it is fitted with controls situated inside the car or within reach of a person inside
- Means of transport of persons using rack-and-pinion rail mounted vehicles
- Mine winding gear

- Theater elevators
- Construction site hoists intended for lifting persons or persons and goods

2.4 MACHINES WITH AN INCREASED RISK

There is a category of products which, because of their nature, are so dangerous that the legislators have decided that they must not be certified autonomously by the producer or importer. An external inspection body, a so-called "Notified Body," must be engaged for this. This also applies to certain categories of safety components. Appendix IV gives a list of useful addresses in Europe. A summary of machines with an increased risk is included in Annex IV of the Machinery Directive. This Directive can be found in Appendix I.

If Annex IV is applicable, a distinction must be made between a machine manufactured in accordance with harmonized standards and a machine which meets the requirements of the Directive in some other way. This distinction is logical as only the use of harmonized standards gives a presumption of conformity.

If a machine which falls under Annex IV has not been produced in accordance with harmonized standards, the manufacturer must provide a Notified Body with a model in order for an EC type examination to be carried out.

A manufacturer who has produced the product taking account of the harmonized standards has the following options:

1. The manufacturer can personally draw up a Technical Construction File and send it to a Notified Body. This Notified Body will then send an acknowledgment of receipt and keep the Technical Construction File (see Annex V, Point 4a, of the Machinery Directive).
2. The manufacturer can opt to have the Technical Construction File submitted for checking by the Notified Body. This check will consist of the Notified Body verifying the correct application of the harmonized standards. If the Notified Body is of the opinion that the harmonized standards have been applied correctly, it shall send a declaration of suitability to the manufacturer (see Annex V, Point 4c, second line, of the Machinery Directive).

3. If desired, the manufacturer can still submit the machine to an EC type examination (see Annex V, Point 4c, last line, of the Machinery Directive). A manufacturer who wants to be sure that a machine meets the requirements of the Directive shall choose the last option. The difference from the aforementioned option is that in this option the machine is checked by the Notified Body, while in the previous option only the files are sent.

2.5 RELATIONS WITH OTHER DIRECTIVES

Article 1, Section 4, of the Machinery Directive states:

> If, for a machine or a safety component, the risks specified in this Directive fall fully or partly under a specific Community Directive, the present Directive shall not or no longer apply to those machines or those safety components and those risks as soon as the specific Directive is applied.

The Directives cited in this exception replace the Machinery Directive only for the risks concerned. The Machinery Directive withdraws as regards that special risk and allows the specific Directive to prevail.

The following Directives may be of complementary applicability, that is, applicable to a certain risk of a machine falling under the Machinery Directive:

- Directive 87/404/EEC on simple pressure vessels
- Directive 89/336/EEC on electromagnetic compatibility (EMC)
- Directive 90/396/EEC on appliances burning gaseous fuels
- Directive 92/42/EEC on new hot-water boilers fired by oil or gaseous fuels
- Directive 73/23/EEC on the safety of electrical material (Low-Voltage Directive)
- Directive 94/09/EEC on equipment and protective systems intended for use in potentially explosive atmospheres
- Directive 95/63/EEC on safety and health requirements for the use of work equipment by workers at work

The latter, the Directive on safety and health requirements for the use of work equipment, is not a product Directive, but a Directive on the safety of working with work equipment, including machines, geared to the professional user.

2.6 STANDARDS

The Directives which fall under the New Approach are elaborated by means of standards. A standard is a formulated criterion for reaching unity in an area where differences are inappropriate or unnecessary. Standards which have been accepted at a European level are marked by the prefix "EN." A national standard also gets a prefix which is different for every country within the EEA. For example, the prefix "NEN" indicates that this standard has been converted into a Dutch version. The prefix "pr" means that the standard has not yet been definitively approved as a European standard. In practice this is then just a formality. The standard can indeed be used already because the risk of amendments is extremely small. If a machine is designed and produced exclusively in accordance with the harmonized standards, this automatically leads to the presumption of conformity with the Machinery Directive.

The number of standards relating to the safety of machines is too great to cover briefly in this section. Out of necessity this book is therefore restricted to the most general and elementary standard on the safety of machines. For the sake of orientation, the appendices to this book contain a list with a brief description of the major (European) standards which have been formulated in connection with the Machinery Directive.

Three types of standards, diagrammed in Figure 2.1, can be distinguished in the Machinery Directive: type A standards, type B standards, and type C standards.

2.6.1 Type A standards

Type A standards are general in nature. They cover general safety aspects and can be applied to all machines. They form the basic tools for every designer/constructor. A description of some important type A standards for machines follows.

EN 292-1. This standard relates to the safety of machines and covers the following subjects:

- Area of application
- Explanation of basic terms such as danger, safety functions, and mobile protection
- Reference to other standards such as EN 292-2 and EN 1050
- Description of dangers caused by machines such as mechanical danger, electrical danger, and thermal danger
- Strategy for the selection of safety measures
- Risk assessment with a reference to EN 1050

Appendix B of this standard includes an alphabetical list of keywords with a reference to the articles in EN 292-1 and EN 292-2 in which they are explained.

EN 292-2. This standard covers the general approach for reducing the risks of machines in order to increase the safety of the machine. This standard covers the following subjects:

- Area of application
- Reference to other standards such as EN 292-1 and EN 294
- Reduction of risk through design by eliminating or reducing as many potential dangers as possible by means of a suitable choice of design features such as avoiding sharp edges and corners and avoiding electrical danger
- Reduction of risk through design by limiting the exposure of persons to danger
- Safeguards such as the choice of protection and safety provisions
- Information for the user such as instructions, drawings, and written warnings
- Additional precautionary measures with a view to emergency situations such as emergency stop provisions

EN 414. This standard gives rules for presenting and drawing up safety standards. The standard covers the following subjects:

- Area of application
- References to other standards such as EN 292-2 and ENV 1070
- Explanation of type A, B, and C standards
- General principles
- Preparation for drawing up a safety standard such as defining the subject, area of application, and establishing dangers and hazardous situations

Appendix A of the standard can serve as a check list for identifying possible dangers which are applicable to a machine to be designed or assessed. The appendix gives a list of dangers which refers to the relevant paragraphs in EN 292-1, EN 292-2, Annex I of the Machinery Directive, and also some type B standards.

EN 1050. This standard covers, among other things, the method for quantitative risk assessment. This standard also gives a brief discussion on methods of identifying risks in the different use situations of machines.

ENV 1070. To a large extent this standard relates to the naming of the aspects which relate to the safety of machines. In order to prevent confusion and misunderstandings, it is desirable to use unambiguous terms.

2.6.2 Type B standards

Type B standards cover specific technical safety aspects and provision and can be universally used for all machines. A distinction is also made between B1 and B2 standards.

B1 standards give a more detailed explanation of safety principles described generally in A standards. Examples are EN 294 (Safety distances for preventing hazardous zones being reached by the upper limbs) and EN 349 (Minimum distances for preventing injury to human body parts).

B2 standards are a more detailed explanation in technical solutions of safety principles described in A standards. Examples are EN 574 (Two-handed operation) and EN 418 (Emergency stop equipment).

To date, the following categories of safety aspects have been addressed by these standards:

- Machine safety provisions, for example, minimum distances to avoid injury to the human body
- Electrical equipment in industrial machines
- Ergonomics: general
- Ergonomics: anthropometry (measurements in relation to the human body)
- Ergonomics: physical loads
- Ergonomics of the physical working environment
- Vibrations
- Acoustics

Below are some descriptions of important standards.

EN 60204-1. The standard EN 60204-1 gives general requirements relating to the electrical and electronic equipping of industrial machines. This standard gives a so-called presumption of conformity with the Low-Voltage Directive (73/23/EEC). The standard is very important for machinery construction because dangers connected with electricity must also be avoided in an electrically powered machine. In particular, it covers the following matters:

- Reference to other standards such as EN 292-1 and 292-2
- Area of application
- Definitions such as "defective," "housing," and "equipment"
- Description of possible dangers such as fire, electric shock, and interruptions
- Protection against the danger of contact with electricity
- Protection of equipment
- Switching equipment such as position, assembly, and housing
- Technical documentation

EN 294. The standard EN 294 relates to the safety distances to prevent upper limbs from reaching hazardous zones. This standard covers the following subjects:

- Area of application
- Reference to other standards such as EN 292-1
- Definitions of protective structure and safety distance
- Values for safety distances
- Effect of extra protective constructions relating to safety distances

EN 349. The standard EN 349 relates to the minimum distances for preventing injury to human body parts. In particular, this standard covers the following subjects:

- Area of application
- Reference to other standards such as EN 292-1, EN 292-2, and EN 294
- Definition of jamming zone
- Description of minimum distances

Appendix A to this standard gives an example of jamming zones.

2.6.3 Type C standards

Type C standards give safety specifications for specific (groups of) machines. In particular, these standards are generally an excellent aid for a designer/constructor for meeting the fundamental health and safety requirements of the Machinery Directive. It is therefore advisable to examine whether a C standard has been drawn up for a machine; the application of such a standard automatically gives presumption of conformity.

The following categories of type C standards can be distinguished:

- Agricultural machinery and implements
- Building machines
- Conveyors
- Earth-moving machines
- Food-preparation machines (machines for processing)
- Heat-treatment technology
- Leather, synthetic leather, and footwear machines
- Machine tools
- Metalworking machines

- Packaging
- Pumps
- Print and paper-processing machines
- Refrigerating machines (industrial)
- Rubber and plastics machines
- Timber-processing machines
- Transportation machinery (mobile)
- Transportation machinery (robots)
- Transportation: hoisting mechanisms and lifting devices

Some examples of type C standards are:

EN 500-1

This standard contains general requirements for mobile road-building machines.

EN 500-2 to EN 500-6

These standards contain specific requirements for mobile road-building machines.

EN 536 draft

This standard contains requirements for machines for producing building materials.

EN 792-1 draft to EN 792-4 draft

These standards contain requirements for hand tools with nonelectric power.

2.7 RESPONSIBILITIES

The Machinery Directive places responsibilities on specific (legal) persons. The (legal) persons who are made responsible for affixing the CE Marking are:

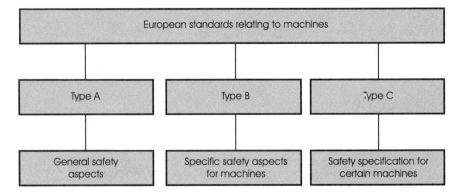

FIGURE 2.1 European standards in relation to the Machinery Directive

- The manufacturer within the EEA
- The authorized representative of the manufacturer within the EEA
- The importer within the EEA

(See Figure 2.2.) Note that the CE Marking is affixed only to complete machines; the manufacturer or supplier of machine parts or semi-finished products has a different set of responsibilities.

In principle the manufacturer is the person appointed to comply with the obligations set by the Machinery Directive. If the manufacturer is not based within the EEA, these obligations lie on the authorized representative or the importer. In this way someone is always liable for noncom-

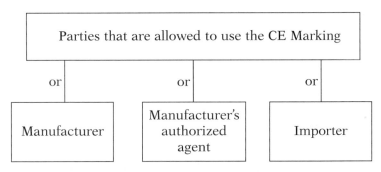

FIGURE 2.2 Who affixes the CE Marking?

pliance with this responsibility. In these cases criminal sanctions may be imposed.

Examples are a sales prohibition on the machine and/or a hefty fine. It is also possible for changes to be imposed on machines already on sale or about to be launched.

2.7.1 Manufacturer of complete machines

The manufacturer is the party responsible for the design and production of a machine. Therefore the manufacturer is responsible for the procedures imposed by the Machinery Directive. As long as the major tasks are being carried out by the manufacturer, such as designing and producing, this party should affix the CE Marking.

Outsourcing by the manufacturer. If certain tasks are outsourced, such as the design and/or production of complete machines, the party that is identified as the manufacturer may vary. If the production of machines should be executed in accordance with the precise specifications and design drawings drawn up by the purchaser, the *contractor* has little input into the way in which safety is implemented in the machine. In fact, only workforce and manufacturing resources are then brought in. It goes without saying that the purchaser should be indicated as the manufacturer.

On the other hand, if the purchaser only orders a certain type of machine without entering into detail, the contractor has a completely free hand in manufacturing the machine. In this case the contractor has influence on compliance with the safety requirements and should therefore be indicated as the manufacturer.

If the manufacturer engages an engineering bureau for the design of a machine, the engineering bureau cannot be labelled as the manufacturer. The bureau has no interest in the production and sale of the machine. The bureau is engaged for expertise in the field of designing. It may be expected that the engineering bureau will integrate the Machinery Directive in the design.

On the other hand, it is the case that when a design bureau commissions a manufacturer to produce the designed machine, the responsibilities lie with the design bureau.

Launching machines on the market. It is important that the term "launch on the market" is interpreted correctly, and that its applicability to a product already in stock and then launched on the market is understood. The definition used by the European Commission for the term "bring into circulation," which means the same as launch on the market, reads as follows:

> The first making available, against payment or free of charge, of a product covered by the Directive in the Community market for the purpose of distribution and/or use on the Community territory

The definition in fact says that machines delivered from stock after the transition period of the Machinery Directive (1 January 1995) *should meet* the formalities imposed by this Directive. This applies both for producers and for importers in the EEA. There is no exception of applicability for machines which are produced or sold in an unchanged form. What is fully applicable is that the machines are being newly launched on the European market and must therefore have the CE Marking.

Machines in stock. A stock is a quantity of products which is present in the distribution chain for sale. It makes no difference to this definition whether the products are on shelves in the warehouse or on shop shelves. Those who have the machines in stock must in some cases comply with the formalities imposed (see Chapter 3) by the Machinery Directive. Whether this is actually the case depends on the time the product was first launched on the market.

If the machine was launched on the market *before* the end of the transition period of the Machinery Directive (1 January 1995), the formalities do not need to be met. This exemption applies not only to machines which were already with the end user, but also to machines that were in stock at wholesalers or retailers.

If the product was launched *after* the transition period of the Machinery Directive, the imposed formalities must be complied with. Therefore, machines which are at the end user must have the CE Marking. Whether machines which are in stock after the enforcement date should be certified depends on with whom and when they entered stock.

2.7.2 Manufacturer of machine parts and semi-finished products

No CE Marking can or may be affixed to machine parts. Indeed, when parts are supplied it cannot yet be foreseen how they will be used in the complete machine by the user. The manufacturer of parts therefore has no responsibility for adhering to the formalities imposed by the Directive.

Of course, this does not mean that the manufacturer could deliver an unsafe machine part. A manufacturer of parts must indicate how the part must be installed and for what purposes the part is suitable. By supplying a IIB declaration (see Appendix I, Annex IIB) with the part, the manufacturer declares that the part concerned may only be applied and used if the machine in which the part is incorporated has a CE Marking. This IIB declaration is also not obligatory. It is advisable to supply a IIB declaration because the manufacturer can limit liability with it. In this way parts manufacturers indicate that they are responsible for the proper functioning of their parts only if they are used in a machine which meets minimum safety requirements. A manufacturer of a machine part is therefore not liable if a complete machine breaks down unless the defect is entirely attributable to the supplied part.

2.7.3 Authorized representative of the manufacturer

A person who has concluded an agreement with the manufacturer outside the Community to act within the Community as the manufacturer in the areas on which the two parties have agreed may be appointed as the authorized representative of the manufacturer. For example, this may be a subsidiary company with the same name as the parent company. The authorized representative can act as the manufacturer only in those areas that are permitted by the Directive and on which there is agreement with the manufacturer. These mostly relate to the administrative obligations of the manufacturer.

The authorized representative is not authorized to modify the machine. In Annex I of the Machinery Directive the authorized representative is only named in the writing of the user's instructions. Furthermore, the authorized representative has no authority as regards the essential requirements of the Directive. If a manufacturer transfers the fulfilment of administrative obligations to an authorized representative, then the latter should be based within the Community.

The authorized representative must be based within the EEA. If the manufacturer has not done so, the authorized representative must affix the CE Marking because representatives are liable as legal persons under the legislation of the member state in which they are based. Settlement of claims for damages between the authorized representative and the parent company fall outside the specifications of the Directive.

2.7.4 Importer of machines

Importers are responsible for machines they import from outside the EEA. It is therefore important for importers to make good arrangements with their suppliers. Used machines which are imported from outside the EEA are also treated the same as new machines and therefore must also be certified.

If the machines are obtained from one of the member states of the EEA, the importer is not responsible for affixing the CE Marking. Whether the importer has to comply with the imposed formalities depends on the time of launching on the market.

Launch on the market before the end of the transition period. If a secondhand (used) machine was launched on the market *before* 1 January 1995, the imposed formalities do not need to be complied with. After all, compliance with the Directive was not yet a legal requirement at that time.

Launch on the market after the transition period. If a secondhand (used) machine was launched *after* this date, it also does not (in principle) need to comply with the imposed formalities because the machine was already on the market within the EEA. For example, if a machine was sold in Germany after 1 January 1995, the machine will not need to be certified again if the machine is launched on the market for a second time, for example, in the Netherlands. However, some exceptions do apply:

1. As of 1 January 1997 machines also need to comply with the requirements of the Directive on safety and health requirements for the use of

work equipment. This Directive does not state that the CE Marking needs to be affixed, but does impose other formalities which agree with those of the Machinery Directive.

2. If the function of the secondhand machine is changed before it is launched on the market again, then after modification of the machine, the risks of the product should be examined again and the CE Marking affixed (again).

3. An importer who launches a machine on the market must take account of (product) liability legislation. It may not be necessary to comply with the formalities of the Machinery Directive, but other obligations arize from the Directive on product liability.

In practice, the above situation can result in necessary problems. The importer has no technical files, and the user's manuals are also often in the language of origin of the product. It is quite possible that the producer based outside the EEA still ensures that the machine complies with European legislation and standards. These obligations for submitting files and writing user's manuals can be laid down contractually between the importer and the producer outside the EEA.

Import of machine parts and semi-finished products. The same situation applies to the importer as to the manufacturer (see Section 2.7.2).

2.8 ASSEMBLY OF SEMI-FINISHED PRODUCTS

Assuming a company is domiciled within the EEA and assembles products based on supplied semi-finished products, the following situations may occur.

Assembly on the company's own initiative

The company assembles products according to the objectives of the applicable Directive(s) and is itself ultimately responsible. It then affixes the CE Marking itself according to the applicable Directive(s).

Assembly under commission

The company is acting under commission of, for example, a firm of engineering consultants who have also drawn up the design. The assembly company supplies, as it were, a service to its commissioning client and is therefore not ultimately responsible. The firm of engineering consultants here is the one that is to affix the CE Marking according to the applicable Directive(s).

The extent to which the assembly company can be held responsible by the commissioning client for damage resulting from an unforeseen accident depends on the written agreements made in the client's commission to the assembly company. A very important aspect here is that watertight agreements have been reached regarding, for example, supply conditions.

2.9 PRIVATE LABEL

Some commercial companies apply their own trade names to purchased products. This is known as a "Private Label." In this way, they conceal the original manufacturer from the user. Here, the legal entity representing its own trade name must affix the CE Marking according to the applicable directives.

If this concerns products produced by manufacturers within the EEA, then these products must already bear the CE Marking. Legal agreements between the producer and the trading company can determine how to proceed in the case of claims for compensation, but in the eyes of the user, the owner of the trade name is the person who can be held responsible.

2.10 INSTALLATION OF MACHINES

Installation of a product or part is the application or setting up and preparation for use of this product or part. The person installing a product or part must in some cases comply with the formalities imposed by the Machinery Directive. Whether compliance is actually required depends on exactly what is installed (full configuration, different pieces of equipment, etc.), where the products or parts come from, and whether the installer also carries out specific activities.

Configuration complies with CE Marking

Installers will always have a preference for installing configurations which already have the CE Marking. They have no responsibility for carrying out the formalities imposed by the New Approach Directives if they install products in the same condition as received from the producer. In this case the producer is fully responsible for making the product comply with the Directives and should affix the CE Marking and also supply the correct installation instructions with the product. It goes without saying that the installer must observe the instructions. Therefore it is important that the instructions are archived so that they can be used as reference and/or covering material in the event of being held liable.

Different parts comply with the CE Marking

It is possible that the installer must install parts which each separately have the CE Marking. This refers to different parts (of different products) which have been certified individually. The producers of those different parts must comply with the formalities imposed by the New Approach Directives (including the installation instructions). The composition of the marked parts will also comply with the CE Marking. The installer should accurately observe and archive the installation instructions (from the different producers) in this case too. In this way the installation is seen as a final phase of production. In this scenario installers are not responsible for certifying parts, but they are held responsible for the CE Marking of the complete product because they are the first point of contact for users.

Various parts without CE Marking

In contrast to the above, the installer may also install parts which do not have the CE Marking. An example of this is if the installer provides a system or device for the installation. In this case the installer is equated to the producer and is therefore responsible for the CE Marking of the complete product. It is important here that the producer of the noncertified parts still send installation instructions, even if this is not strictly an obligation arising from the New Approach Directives.

Installer imports from outside the EEA

If the installer imports parts or entire products from outside the EEA, these parts or products will not have the CE Marking. The installer as the importer is also equated to the producer in this case and is therefore responsible for certifying the product. It is important here that the producer of parts or entire products send installation instructions, even if not obliged to as a producer outside the EEA.

Installer makes modifications

A specific activity of installers which often arises is the modification of existing products. When installers repair or maintain existing products at their own discretion, there is nothing wrong within the framework of the CE Marking provided that they replace existing parts with identical ones. If an installer makes functional or elementary changes to a product during installation, new features may arise which invalidate the existing CE Marking. These features should be examined again. In other words, after a functional or elementary change, the installer who made this modification is responsible for the formalities arising from the New Approach Directives.

2.11 HIRING OF MACHINES

A person who hires out a machine must in some cases comply with the formalities imposed by the Machinery Directive. Whether this is actually the case depends on the time of hiring out.

Hiring before the end of the transition period of the Machinery Directive

If the machine was hired out before this date (1 January 1995), the imposed formalities did not have to be complied with. After all, compliance with the Directive was not yet a legal requirement. However, other existing statutory regulations had to be observed because with hiring out, a product is always being put into circulation again. The regulations are as follows:

1. User's manual. When it is hired, every machine should be accompanied by a correct user's manual (see Section 3.4 in Chapter 3). This manual (the user's instructions, safety regulations, and other regulations) can be legally defined as a set of (obligatory) rules of conduct for the hirer.
2. Periodic and recorded maintenance. Periodic and recorded maintenance is of the utmost importance for hiring out machines: the hire company should guarantee continued safety of a machine when repeatedly offering it for hire. Maintenance should take place periodically and be recorded. The maintenance carried out must be put in writing. In this way the condition in which the machine is being circulated is clear as well as whether it still complies with the safety requirements.

These exceptions are connected to the fact that the hire company cannot supervise the use of the machine. For example, it may be that a previous customer has dismantled the protective cover of a dangerous part. Without the protective cover, the machine no longer complies with the requirements imposed by the Machinery Directive, even though a CE Marking has been affixed.

Hiring after the transition period of the Machinery Directive

For machines hired out *after* 1 January 1995, the imposed formalities must (in principle) be complied with. It makes no difference here whether the machine was produced within or outside the EEA.

2.12 THE SALE OF MACHINES WITH A CE MARKING OUTSIDE THE EEA

As the word suggests, European legislation is only a legal obligation for countries within the European Economic Area. If machines are exported to a country outside the EEA, the CE Marking is not a legal obligation and the national measures of the country concerned must be complied with. Everyone is free to export machines with or without a CE Marking to a location outside the EEA.

If a trading company has a range of machines intended for sale both within the EEA and outside, it is advisable for all the machines to be brought into line with the European legislation. This prevents confusion and guarantees a machine which complies with the essential requirements relating to health, safety, the environment, and consumer protection.

3
Procedure for CE Marking under the Machinery Directive

3.1 INTRODUCTION

For every Directive a number of formalities must be complied with before the CE Marking may be affixed to the product. These formalities can differ depending on the Directive and the product. Overall, however, the following formalities are applicable:

- Carry out a risk assessment.
- Put together and archive a Technical Construction File.
- Write a user's manual.
- Apply for and archive an EC type declaration.
- Draw up and sign the EC Declaration of Conformity.
- Affix the CE Marking.

3.2 RISK ASSESSMENT

Besides being the basis for a safe product, the risk assessment also forms the starting point for the user's manual and the Technical Construction File (TCF), which will also be described in this chapter. By gathering as much knowledge as possible about the user and the circumstances of use and combining these, it is possible to reach a safety assessment of any situation which may occur. In this chapter it will become clear how you can recognize and assess risky situations. Acting on this, risks should be eliminated as far as possible and safety measures should be taken.

3.2.1 Introduction

It is often not possible for designers to eliminate or reduce all the risks associated with a product. They will have to make well-founded choices when establishing which risks they will approach in the first instance. The risk assessment is an aid to setting priorities. The designer must be able to substantiate those decisions relating to product safety. It is therefore very important for the risk assessment to be carried out systematically. This principle is the basis for self-certification and the TCF.

The risk assessment itself and taking account of the results arising from it takes some time. Indeed, a major change has taken place in the importance of both. A systematic risk assessment on the one hand and good documentation of the design decisions relating to safety measures on the other hand are very important because a designer must be able to legally defend design decisions as regards safety aspects. In the case of the Machinery Directive, producers themselves are obliged to carry out a systematic risk assessment and to keep documentation for at least ten years after production.

Various concrete targets can be achieved by carrying out a risk assessment:

- Discovering risks
- Quantifying risks
- Assessing the acceptability of the risk
- Substantiating the choice of safety risks

The Directive makes no mention of the phase in the design process in which the risk assessment must be carried out. However, the risk assessment must be carried out before the design is completed because the results of the assessment must be incorporated into the design.

Design stage. In the course of the product design process, the layout and form gradually become more concrete. A limited risk analysis can even be carried out at an early stage so that the design can be easily changed. The further advanced the design process is, the more concrete will be the results produced by the risk assessment procedure. A risk assessment early on in the design process should in principle be used only to identify risks and to draw up specifications. The identified risks can be analyzed at a later stage.

In this the considerations behind the CE Marking must be borne in mind: manufacturers must be able to demonstrate at all times that they have made the effort to supply their products with a reasonable level of safety.

3.2.2 Definitions

For the sake of clarity, the terms used should be clearly defined. The following standardized definitions should be used where possible:

Damage: Physical injury and/or damage to health or objects
Hazardous event: An event which can cause damage
Safety measure: A means which eliminates a danger or reduces a risk
Remaining risk: The risk which cannot be eliminated by a safety measure

3.2.3 Step plan

To assure the person carrying out the analysis of a systematic approach, the various phases of the risk analysis can be described using a step plan:

Step 1 Identification of risks
Step 2 Assessment of risks
Step 3 Risk evaluation
Step 4 Risk reduction
Step 5 Safety evaluation

3.2.4 Identification of risks

The operation of a machine implies a certain interaction between a user and the product. Furthermore, bystanders, housemates, or neighbors may come into contact with the product. The output includes both desired aspects, characterized by the intended function of the product, and undesirable aspects such as noise, vibrations, and waste.

The interaction comes into being in a specific environment which cannot always be influenced by the designer. The designer will have to take account of three factors: people, product, and environment.

Thus, people-related properties are important in the interaction with the product, which means that not just the physical properties of the user, but also the psychological properties of the user have an influence on the user's manuals. The environment in which the interaction takes place is formed by physical and social factors. Under physical factors consideration must be given to everything which is "touchable" in the environment. Air temperature and humidity are included in this. Thus there is a risk of short-circuiting from the moisture in the bathroom and there is danger if the user does not hear a warning signal because of the noise of passing cars. Under social factors, where the background of the user plays a part, consideration may be given to the level of training which may affect the handling of the machine or the ability to understand the user's manual. The features of the product may also affect the product safety. Some product-oriented examples of dangers are:

- Mechanical dangers
- Electricity
- Temperature
- Radiation
- Fire and explosion
- Chemical dangers
- Power supply

Check lists. Identification of latent risks can be done well by using check lists, provided that they are used correctly. Check lists contain a summary of points for consideration relating to latent risks which could be present in the machine. By using these check lists it is possible to examine item by item what factors are relevant to product safety. It should be noted that the check lists never completely cover everything in every case. Check lists can be found, among other things, in the harmonized standard EN 414, but of course it is also possible to draw up an adequate check list yourself. Appendix III contains a number of check lists which give a good indication of how you might set to work.

3.2.5 Assessment of the risks

Once the risks have been identified, a risk assessment should be carried out for each individual risk using risk factors. The risk entailed in a specific situation or technical process is a function of the following factors:

- The scale of the damage
- The risk of the damage occurring
    (a) Frequency and duration of exposure of people to the danger
    (b) The risk of a hazardous situation occurring
    (c) The technical and human possibilities for avoiding or limiting damage

There are various methods available for the systematic analysis of these factors. These are partially described in the standard NEN-EN 1050.

A specific description of these methods can be found in the practical manual "Risk analysis and user's manual for European Directives" which you can also obtain from Kluwer Techniek in the European Directives series.

3.2.6 Evaluation of the risks

In the risk evaluation it must be established which risks are acceptable and which are unacceptable. This activity cannot be carried out objectively, as the objectivity is dependent on factors such as the personal experience of the person carrying out the evaluation, the social acceptability of the risk concerned, and the desired image of the company.

Account should be taken of the fact that a reasonable margin must be used because of the unavoidable subjectivity in the execution of the risk evaluation. In this respect it is recommended that comparable products and the safety level that they offer be looked at. A second recommendation is to have the risk evaluation carried out by at least two people.

An exact description of an acceptable risk cannot be given. The analyst must now be able to evaluate, using only the information available, which risks are acceptable and which are not.

3.2.7 Risk reduction

The fourth and last phase of the risk evaluation is to reduce the risks. According to the requirements of the Machinery Directive, when choosing the most appropriate solutions, the following principles should be applied in the order given below:

- Eliminate the risks or restrict them as far as possible (in the design and the construction in the safety incorporated into the machine).
- Take the necessary safety measures for risks which cannot be eliminated.
- Inform the user about the risks which are still present as a result of insufficiency of the safety measures taken, indicate whether special training is required, and state that personal protective clothing must be worn.

Achieving risk reduction targets in accordance with EN 1050. By achieving the following conditions, it can be demonstrated that the risk reduction process can be concluded:

- The danger has been eliminated or the risk has been reduced by
    (a) Changing the design or replacement with less hazardous materials and substances
    (b) Technical safety
- The selected safety measures are of a type which have been proven to provide sufficient protection for the intended use.
- The type of safety measure chosen is suitable for the application in terms of:
    (a) The probability of elimination or obviation
    (b) The scale of the possible damage
    (c) Hindering execution of the required task
- The information on the intended use of the machine is sufficiently clear.
- The work procedures for the use of the machine conform to the abilities of the staff using the machine or other people who may be exposed to the dangers associated with the machine.
- The recommended safe work procedures for the use of the machine and the relevant training requirements are described adequately.
- The user has sufficient information about the remaining risks in the various phases of the life of the machine.
- If personal protective clothing is recommended, the need and the training requirements for the use must be adequately described.
- Additional precautionary measures are adequate (see Chapter 6 of standard EN 292-2:1991).

3.2.8 Safety evaluation

The question which will always be asked in legal proceedings relating to product liability is: "Did you consider this alternative?" In this case the producer must be able to answer, "Yes, these are the reasons why the alternative was not chosen." A sound risk analysis can be very helpful to the producer in this.

The producer will have to be able at all times to justify why certain choices were made, and to indicate what risks remained and why they are acceptable. In other words, it is up to the producer to demonstrate that a product being brought onto the market is safe.

3.3 TECHNICAL CONSTRUCTION FILE

The most important formality of most Directives is to draw up the technical construction file (TCF). This file contains the technical basis which must show that all the requirements of the Directive have been complied with. The file here forms part of a separate "route" which must be followed if the Directive is not automatically complied with via harmonized standards.

The technical construction file must be archived for a period of at least ten years effective from the production date of the last unit. The technical construction file must always be able to be submitted to the authorized monitoring body. If there is a suspicion that a product does not meet specific requirements of the Directive, a monitoring body can submit a request, with its reasons, to inspect the relevant part of the technical construction file.

For the manufacturer, the technical construction file forms the technical evidence which enables it to defend liability claims. Therefore it is necessary for the TCF to be drawn up in accordance with the Directives and with care.

A technical construction file will broadly consist of:

- A general description of the product
- Design, manufacturing drawing and diagrams
- Detailed technical data for essential aspects of the product
- List of standard and fundamental requirements which have been particularly observed
- Reporting of calculations and tests carried out
- Certificates and inspection reports
- For mass production, the internal provisions which were observed to be able to guarantee compliance with the Directive

If a manufacturer does not manufacture in accordance with the European harmonized standards, there is no presumption of compliance. In that case the technical construction file must show in what way compliance has been achieved.

3.4 THE USER'S MANUAL

3.4.1 Introduction

Section 1.7.4 of Annex I of the Machinery Directive discusses the requirements with which the user's manual must comply. As discussed earlier in this book, the manual has several explicit objectives. One of these objectives is making clear the limits where the responsibility of the importer stops and where that of the user begins. This limit cannot be drawn at random and is defined fairly accurately by the Machinery Directive.

NOTE: By placing a remark in the user's manual, an importer or the authorized agent domiciled in the EEA of a manufacturer in a non-EEA State cannot automatically back out of his or her responsibility or liability if a safety criterion under the Machinery Directive is concerned that could and should be implemented in the design or construction.

If a user holds an importer or authorized agent domiciled in the EEA responsible for the consequences of an accident with the machine, it is always a court of law that establishes liability.

3.4.2 The language of the user's manual

The user's manual must be supplied in the language of the country in which the machine is being sold or used within the EEA. In addition, it will also be supplied in the original language of the country of the EEA in which the manual was written, so that it can be referred to in case of indistinctness. Consider, for example, an incorrect translation. Anybody may make the translation, but because of the great importance of the user's manual, it is recommended that a proficient technical translator make such translations. The use of language must be unequivocal and clear, and adapted to the target group for which the machine is intended.

The importer may prepare the translation, but this is not necessarily the case. The manufacturer based outside the EEA or the authorized agent in the Community may also prepare the translation. However, accidents resulting from an incorrect translation are attributed to the importer or the manufacturer's authorized agent domiciled in the EEA. This is an extension of the regulations relating to product liability. This is why it is advisable for manufacturers and importers to agree clearly about the translation.

3.4.3 Structure of the user's manual

A well-structured user's manual might contain the following chapters:

- The technical specification and other relevant data
- Safety aspects

- Installation, adjustment, and disassembly
- Instructions for use
- Maintenance and repair
- Environmental aspects
- EC Declaration of Conformity

Importers are, of course, free to adapt this structure as they see fit. However, the structure of the chapters determines the structure and accessibility of the manual. This is improved to an even greater extent by adding an index.

3.4.4 Method and tools

Drawings and pictograms

Working with drawings, pictograms, and tables increases the speed and ease with which a user can absorb information. The person compiling the user's manual must be aware that users generally take little effort to read the manual exhaustively before using the equipment. Graphical information, like drawings and pictograms, increases the accessibility and the likelihood that the information is actually absorbed.

Warnings and symbols

If dangerous machinery is involved, it is recommended to print a clear warning on the front of the manual containing the urgent advice to read this manual first before proceeding to install, adjust, operate, or maintain the machine. For safety's sake, this warning can also be printed on the packaging.

By marking important passages of text or warning in the manual (for example, where the safety of the individual is a critical aspect) with symbols, the reader's attention is rapidly drawn to this important information.

Repetitions

The safety section will have already indicated certain general warnings relating to the use of the product. Repeating these warnings when discussing each respective aspect is desirable.

Subjects will often be discussed more than once, for example, when identifying tools to be connected to the equipment for the technical data with an indication that other tools may not be used. This is repeated in the safety section, to the effect that only the listed, prescribed tools can and may be used and that the use of all other tools is advised against. Ignoring a use that has been advised against is therefore the user's responsibility. After all, the importer provides a suitable solution for a reasonably expected use in the form of the prescribed accessories.

Remark When identifying the risks with the aid of the check list(s) in Appendix III, several points of attention can already be entered in the appropriate inventory form. This inventory not only contains references to the user's manual, but also presents subjects that, when viewed more closely, are worthwhile to indicate in the manual.

3.4.5 Technical specification and other technical data

The following should be indicated in the user's manual under the heading "Technical specification and other technical data" (if applicable):

- The information on the EC Declaration of Conformity, except the serial number of the machine
- The technical data of the machine with relation to its use, for example, capacity, weight, dimensions
- The permissible maxima and/or the safety or operating coefficient in use
- Technical conditions for connecting the equipment (for example, a shockproof wall outlet)
- Special environmental conditions
- Data on the noise emission
- Information regarding vibration (in the case of hand-held and/or hand-guided machinery)
- Summary of all the associated components delivered with the equipment and/or accessories and/or machine tools
- Other tools to be connected or used
- If required, a brief introduction to the structure of the machine

3.4.6 Safety aspects

This section is divided into two groups: one part on the safety measures that have to be taken, and another on use that is advised against.

Safety measures to be taken. Often, safety measures focus on the use of personal protective equipment by the user, such as hearing protection, sight protection (dust and light), or protection of the respiratory system. Also considered in this category is the use of gloves or safety shoes.

Attention must also be paid to the design of the operating environment. In that context, adequate lighting and ventilation or air extraction are also very important.

If necessary, statements must be included on special technical aspects, such as permissible tolerances, permissible tools, cutting out when not in use for a certain period, or switching off or removing the energy supply after use.

Use to be advised against. By excluding an incorrect or undesirable application, the importer can never be accused of neglecting to indicate that the machine is not intended for that particular use.

A classic example is a lawsuit that is reported to have taken place in America involving a manufacturer of microwave ovens. The manufacturer had omitted to indicate in the user's manual that the operation of the oven was based on a principle that constitutes a hazard under certain conditions to the health of humans and animals. A pet died after a consumer had placed it in the oven to dry. It is reported that the manufacturer was held responsible and was obliged to pay considerable compensation.

A summary of uses to be discouraged can never be complete. This is why the intended use must always be described explicitly and correctly. In the previous example, therefore, the manufacturer in question should have clearly stated in the user's manual that the device is intended solely for heating foodstuffs.

In the user's manual, it is generally advisable to warn against any use for other applications or objectives than the one for which the machine is intended. The same is true of the application or connection of other tools, components, or machine accessories than the ones prescribed. The same holds true for using the machine in a different environment or with other objects or raw materials than those for which the machine is intended.

It is also advisable to warn against the use of the wrong maintenance materials (cleaning, lubricating, etc.). Use by children must be advised against or even prohibited.

Precisely defined procedures are often available for maintenance and defects. A warning should be issued against deviating from the necessary prescribed procedures. Also consider warnings relating to changing the accessories.

Dangerous or poisonous substances. Machinery may contain hazardous or poisonous substances in order for it to operate, or it may be intended for processing such substances. Information on handling poisonous substances in relation to the Machinery Directive is scanty.

Article 1, Section 3, excludes reservoirs for the storage and pipes for the transportation of petroleum, gasoline, flammable liquids, and dangerous substances from the operation of the Machinery Directive.

Section 1.6.5 in Annex I of the Machinery Directive formulates requirements relating to cleaning or maintaining the internal parts of the machine that contain, or that have contained, dangerous substances or preparations.

The handling of dangerous substances can, of course, also mean dangers for the health of the user. Supplementing the user's manual with information on the handling of dangerous substances and the associated dangers is advisable. Prescribe at least the use of the correct protective materials. Provide the necessary information on the storage of the dangerous substances and on symptoms associated with incorrect use. Provide regulations relating to the presence of first-aid materials and calling on medical assistance in case of accident.

3.4.7 Transport, installation, taking into operation, and dismantling

All the attention is usually paid to the actual use or application of the machine. Yet this is also preceded and followed by certain phases which may very easily be omitted inadvertently. If we list all the elements of a machine's life cycle, we obtain the following summary:

- Transportation
- Installation
- Taking into operation
- Regular operation
- Maintenance and repair
- Dismantling

The subjects in this section relate to all the elements in the machine's life cycle, except regular operation. Depending on the type of machine, certain elements from its life cycle are not applicable. For example, small and light machinery, such as a portable circular saw, will not come with any special requirements for transportation. Yet if the weight is greater and the dimensions are larger, separate provisions have to be made for the product or packaging. This may include hand grips, crane hooks, and fastening points for accessories for transportation.

Transportation. Provisions are sometimes added to a machine for transportation purposes which actually form part of the machine during transportation and are therefore subject to the Machinery Directive. This may relate to the packaging, but also to the tools for transportation. These provisions are subject to the same requirements with respect to health and safety as the machine itself. The packaging may therefore not have any dangerous sharp edges or protrusions, and the material must comply with the environmental requirements.

The tools for transportation include, for example, hoisting, lifting, and loading equipment. They must be used safely and must comply with the essential health and safety requirements of the Machinery Directive, as set out in Annex I of the Directive—in particular, in Chapter 4 of this annex, "Essential health and safety requirements to offset the particular hazards due to a lifting operation."

It is advisable to pay extra attention to the desired position during transportation, the sensitivity to shocks or vibrations, and auxiliary provisions such as supporting points or crane hooks. Also consider the instructions for unpacking the equipment and include warnings relating to critical or vulnerable parts of the machine.

The user's manual will not be read by the shipping company transporting the machine, which will have to make do with the information printed on the packaging. As transportation often takes the machinery across national borders, account will have to be taken of the language differences. This is why it is advisable to use as many pictograms as possible.

Installation. In the installation of a machine, requirements are set for the installation engineer, the environment in which the machine is to be located, and the tools or other instruments required for this. The machine must be designed so that the activities relating to the installation can be conducted safely. The user's manual must pay sufficient attention to the requisite safety measures.

Taking note of the prescribed tools and/or instruments is important. Will it be necessary to supply them along with the machine? Requirements are usually also set for the basis and durability of the fasteners. The same is true o up the working environment—railings, platforms, channels, and the like.

A few other things are very easily forgotten: transportation brackets and blocking components that have to be removed before putting the machine into operation.

Placing the machine into operation. It is self-evident that every machine is placed into operation for the first time only once. The customary activities are setting up the equipment and adjusting it. Points of attention here are the requisite tools, such as implements and measuring devices. Take note also of the evolution of gases or vapors, originating from parts that are being heated for the first time.

Dismantling. When a machine is sold, or at the end of its service life, it will be dismantled. This may be accompanied by the same kind of hazards as during transportation, installation, and operation. If a machine is intended for the scrap heap, we should ask ourselves whether the machine might contain substances that are hazardous to the environment or that form a danger during the scrapping process (for example, petroleum in fuel tanks). It goes without saying that such substances must be removed from the machine promptly and must be disposed of in the correct way.

3.4.8 Instructions for use

Each machine, of course, has its own specific operating method. This information must be conveyed to the user via the instructions for use. These indicate in principle everything that is necessary to operate the machine correctly. A description of the control units is necessary in each case. A description of the emergency-stop facility is also important. Where applicable, the necessary personal protective materials must also be described. A specific education or training course is also often required to operate the machine. This must of course be indicated accordingly.

3.4.9 Maintenance and repairs

Maintenance and repair are necessary, but also involve dangers from a safety point of view. Indeed, accidents occur all too easily in these kinds of situations, for example, because safety provisions have been switched off or because the regulations are deliberately ignored when something is being repaired quickly. However, the supplier clearly has no control over how the rules and regulations are put into practice and therefore must at least indicate the correct method and the necessary information. This may be in the form of diagrams, drawings, instructions, measurement data, adjustment data, or warnings, for example.

The safety of a machine is determined to a significant extent by the expertise and promptness with which the maintenance or the repairs are performed. This is why it is important to include in the user's manual that repairs and possible maintenance may be performed only by an expert.

Maintenance. Maintenance includes the obvious activities such as cleaning and lubricating. General statements like "the machine must be lubricated regularly," however, are out of place in a good maintenance procedure. All activities must be well quantified and qualified. This means *where, by whom, when, how frequently, using what,* and *how much.*

It goes without saying that this is essential for more complicated activities like replacing components or tools. If strict regulations exist for the choice of materials to be used, the use of other materials must be advised against. If substances are involved that are dangerous to health and possibly also to the environment, additional information should be provided.

Repairs. Repairs may be carried out only by experts, even if only for reasons of safety. Carrying out one's own repairs without having the specialized knowledge required for such activities can easily lead to accidents, with all the associated serious consequences, including also the legal consequences.

The following information is important with respect to repairs:

- Addresses of recognized institutes able to repair the machine
- Addresses of suppliers of recommended components
- Warning against possible dangers
- Technical information, such as drawings, diagrams and parts lists

Keeping a logbook in which undesirable phenomena are recorded with relation to the operation of the machine may identify errors at an early stage. This may increase safety in some situations. In addition, such a logbook may help in tracing "tricky" malfunctions.

3.4.10 Environment

The Machinery Directive does not set any direct environmental requirements. Today of course, however, the environment can no longer be disregarded when trading and operating machinery. Let's be honest: there is important common ground between health and safety, on the one hand, and the environment, on the other. So it is also desirable from the viewpoint of the Machinery Directive to pay necessary attention to the environmental aspects.

To start with, some machinery uses substances or tools that are hazardous to the environment. Consider oil, batteries, or chemicals. These may be released during maintenance activities or they may have to be removed when the machine is to be dismantled for disposal on the scrap heap. The user's manual must set requirements for all these environmental aspects.

If environmentally dangerous substances are involved that may also be dangerous to health, the user's manual must pay special and separate attention to this subject. Information must be provided on the substances involved and the way in which they are to be handled. It is also advisable to warn against possible symptoms as a consequence of contact with hazardous substances. Another important factor is the measures to be taken in case of accidents involving these substances and, if possible, the methods or procedures for disposing of them.

3.5 EC TYPE DECLARATION

Certain products manufactured in quantity may not be certificated individually by the manufacturer, but might have to be subjected to the EC type examination. This examination involves the inspection of a representative example (type) of the series in question by, or on behalf of, an external inspection organization or Notified Body within the EEA.

The Notified Body assesses whether the type meets the essential criteria in the Directive, conducts tests if necessary, and, after approval, issues a declaration of the EC type examination. The declaration contains the name and address of the manufacturer, the results of the inspection, the conditions for the validity of the declaration, and the most important data through the identification of the approved type. Thus the EC Type Declaration is an official document issued by the Notified Body after approval of the type.

3.6 EC DECLARATIONS ACCORDING TO ANNEX II OF THE MACHINERY DIRECTIVE

According to Annex II of the Machinery Directive, there are three types of declarations; they are explained in Sections A, B, and C of the Annex. This explains the distinction in the names of these three declarations, namely:

- EC Declaration of Conformity IIA
- Manufacturer's Declaration IIB
- EC Declaration of Conformity IIC

EC Declaration IIA is required for all machinery and interchangeable equipment bearing the CE Marking. The Manufacturer's Declaration IIB is intended for all machinery that is unable to operate independently and that is intended to be installed in a greater entity. Declaration IIC is especially intended for safety components placed on the market separately. Incidentally, the CE Marking can and may only be affixed to machinery and interchangeable equipment for which an EC Declaration IIA has been drawn up.

Declarations of Conformity must be added to every machine. They form essentially an important component in the phased schedule leading to CE Marking, but as stated above, the CE Marking may only be affixed to "IIA machinery."

The declarations provide clarity about the identity of the signatory, the importer, or the authorized agent domiciled in the EEA of a manufacturer based in a country outside the EEA. This is the person who indicates that the product in question has been designed and executed according to the Directives indicated in the declaration. At the same time, the declarations also indicate that the user's manual meets the requirements of the Directive.

The declaration in question is supplied with the respective product. This is often one page from the user's manual. The same applies to the language in which the declaration is drawn up as applied to the user's manual. So the declaration must be drawn up in the official language or languages of the country of the EEA where the machine is in use. The declaration must also be drawn up in the language of the country of production.

3.6.1 EC Declaration of Conformity IIA

This document is intended for machinery or machine accessories. Annex II, Point A, of the Machinery Directive explains the declaration in greater detail. Important information to be entered here relates to:

- The identity and signature of the importer or the authorized representative established in the EEA of the USA-based manufacturer
- The identification of the machine
- The date of signature

If the machine was certificated by a Notified Body the following must be added:

- Name, address, and number of this body in the EEA
- Number of the EC type examination
- Reference to the relevant requirements that the machine meets (Directives and standards)

Remark

If the machine is supplied with accessories or other attachments, the EC Declaration of Conformity also applies to these attachments and their operation. In other words, the attachments must also have been designed and manufactured according to the essential safety requirements as set out in the Machinery Directive. A supplement must be added to the user's manual accordingly.

A sample IIA declaration is shown in Figure 3.1.

3.6.2 EC Declaration of Conformity IIB

No CE Marking can or may be affixed to machine components. This relates after all to a component, and it is as yet impossible to see how the user is to employ it in the overall machinery. Of course, this does not mean that the producer would be allowed to supply a machine component that is unsafe.

This declaration is a statement that the machinery must not be put into service until the machinery into which it is to be incorporated has been declared in conformity with the provisions of the Directive. Annex II, Point B, of the EC Directive discusses this EC declaration.

By way of an example, we take a hydraulic cylinder that is used to operate a press. This hydraulically operated press is the ultimate machine to which the CE Marking is to be affixed. The cylinder used here must also meet several requirements, if the press is to work reliably and safely. Here, the requirements for the cylinder relate to, among other things, the utilization factor or the permissible excess pressure. They are identified in specific standards in the field of hydraulics. This is why the Manufacturer's Declaration IIB refers to standards and, of course, preferably, harmonized standards.

Under the Machinery Directive an importer is not obliged to issue a Manufacturer's Declaration IIB, but may be prompted to do so under

EC DECLARATION OF CONFORMITY FOR MACHINES

(Directive 89/392/EEC, Annex II, under A)

We,

Manufacturer: Hedge Inc.
Address: 498 Bush Square, Bethesda, Maryland 20810

Do hereby declare that

ELECTRICAL HEDGE TRIMMER HSW-01

Is in conformity with the essential requirements of the Machinery Directive (Directive 89/392/EEC, as recently amended), especially the relevant requirements for this machine of Annex II of the Machinery Directive.

Level of the sound pressure in accordance with 86/188/EEC and 89/392/EEC, measured by DIN 45635: L_p (sound pressure) < 86 dB (A).

The weighted quadratic average value of the acceleration in accordance with EN 50144: < 2.5 m/s^2.

Is in conformity with the following insulation-requirements:
The machine is insulated by VDE 0740 class II and CEE 20 and therefore can be connected to sockets without grounding.

Drawn up in: Bethesda, Maryland
On: 19 December 1997

Signature:

By: Hedge Inc., R.J. Hedgeman

$C \epsilon$

FIGURE 3.1 The EC Declaration of Conformity

pressure from the customer(s). After all, customers do want to know what they are buying. The IIB Declaration also provides legal cover for the supplier of the component or semi-finished product, since this product may not be put into operation before the customer meets certain requirements.

With respect to the identification of manufacturer and machine component, to standard reference, etc., the contents of Manufacturer's Declaration IIB are very similar to Declaration IIA. The most important difference is the remark:

> It goes without saying that reference can be made only to standards.

3.6.3 EC Declaration of Conformity IIC

Point C of Annex II of the Machinery Directive discusses Declaration IIC in greater detail. This declaration is intended for safety components. They do not bear the CE Marking, but do fall within the scope of influence of the Machinery Directive. Since safety components form part of a greater entity, the CE Marking may not be affixed to them. In this respect, safety components are special machine components.

The Machinery Directive provides little information on the way in which safety components are placed on the market. It is clear that an EC Declaration of Conformity IIC must be drawn up. This corresponds almost precisely to that for machine components. There are, however, several differences. No ban on introduction is indicated, for example. The function must also be described, for example, overpressure protection for a pressure of 20 kPa.

The component must be subjected to an EC type examination by a "Notified Body," if this safety component occurs on the list in Part B of Annex IV of the Machinery Directive. It is mandatory to supply Declaration IIC with a safety component.

Taking account of the liability of the importer or the authorized agent domiciled in the EEA of the USA-based manufacturer, the special implementation of safety components—increasing the level of safety—requires a more comprehensive approach. Besides the formally prescribed data, it is advisable to add extra information—namely, a more detailed description of the component itself and its operation, of the assembly, or of installation regulations—to achieve optimum implementation.

Remark Safety components must always be accompanied by EC Declaration of Conformity IIC. If a safety component is involved that is to be inspected by a Notified Body (see Annex IV of the Machinery Directive), then this must be expressed in the declaration.

3.7 AFFIXING THE CE MARKING

When a certification procedure has been completed successfully, the CE Marking can be affixed. The CE Marking indicates that all the essential requirements under the relevant Directives have been met.

The CE Marking comprises the letters "CE," the graphic design of which is set out in the Directives. The following general conditions also apply in particular:

- If the Directives do not indicate specific dimensions, a minimum size of 5 mm shall be used.
- If the size of the CE Marking is increased or decreased, the proportions of the image must remain intact.
- The CE Marking is affixed to the product or to the identification plate. However, if the nature of the product does not permit this, the CE Marking is affixed to the packaging, the warranty certificate, or the operating instructions.
- The CE Marking must be affixed so that it is visible, legible, and indelible.
- No other marks or symbols may be affixed that may be misleading in respect of the CE Marking.

4

Practical Examples of the
Machinery Directive

4.1 INTRODUCTION

In order to create even more clarity in the long route which has to be followed, this chapter will describe the Machinery Directive in practice. First of all the 23 most frequently asked questions on the Machinery Directive will be answered, after which the CE Marking procedure will be covered using a practical example.

4.2 THE MOST FREQUENTLY ASKED QUESTIONS ON THE MACHINERY DIRECTIVE

Question 1
What are the formalities of the Machinery Directive for the European producer or importer?

First of all a standardized risk assessment must be carried out which must be covered by a hierarchical method. Initially the producer/importer should take note of the relevant safety measures and safety regulations for carrying out a standards examination in order to compare the applicable standards with the product. The producer/importer must then draw up a user's manual in the user's language. This must include the applicable safety precautions. Finally, the producer/importer gathers all the relevant technical information which is required to compile the Technical Construction File (TCF). With the Declaration of Conformity (IIA) the producer/importer confirms that the product complies with the requirements of the Machinery Directive.

Question 2
Why, as a European producer, must I carry out a risk assessment for my drilling machines?

Drilling machines fall into the sphere of operation of the Machinery Directive (89/392/EEC), which came into force on 1 January 1995. One of the legal obligations for the producer is to carry out and document a risk assessment. Using a standardized risk assessment as described in this book, you can establish and demonstrate a safety level for your drilling machine. The risks of the machine which you analyse in this way can be eliminated or reduced by observing technical safety measures and safety regulations.

Question 3
If I possess a user's manual with the product, must I then still carry out a risk assessment? And what does a user's manual have to do with a risk analysis?

It very much depends on the type of product and company whether a risk assessment must be carried out if a user's manual is present. In the event that your company produces products or imports products into the European Economic Area (EEA) and is based within the EEA, depending on the type of product and the Directive which is applicable to it, a standardized risk assessment may be a statutory obligation. If your company acts as an intermediary, then your company simply passes on the original product including the user's manual and packaging in the language of your own country.

Question 4
May I, as a potential purchaser of a machine, inspect the risk assessment before purchase?

In the case of certain products, the producer and importer within the EEA should carry out a standardized risk assessment (see response to previous question). Only inspection bodies such as the Inspectorate of Goods and the Labor Inspectorate are entitled to request and inspect the Technical (Construction) File in which the risk assessment is included. It is not usual for the purchaser to ask the supplier formally to prove the safety of a product. On the other hand, suppliers are prepared to make their Technical (Construction) File completely open.

Question 5
As a (foreign) producer of grill installations, I deliver to many Dutch customers. Must I translate my user's manual into Dutch?

A user's manual should be written in the language of the country in which the product is marketed. In the case of grill installations, that means that your user's manual must be written in Dutch.

Under many New Approach Directives, it is compulsory to provide a translation with the user's instructions. The consideration with this obligation is that safe use of a product is not just determined by the technology. For safe use, maintenance, and repair of the product it is necessary for the user to be instructed clearly and unambiguously. Good instructions contribute to the safety of the product. The risk of accidents is thereby reduced.

In order to reduce translation costs, you may consider using pictograms.

Question 6
What must I include in the user's manual in any case?

A correct user's manual does not just include the purpose of use but also the recommended conditions of use of the product. If, for example, something goes wrong when using vacuum cleaners, the user's manual should always be able to demonstrate that the user was informed correctly.

Example: Imagine you are a supplier of vacuum cleaners. A purchaser of your vacuum cleaner uses it on his or her car in rainy weather. The vacuum cleaner turns out not to be resistant to moisture. If the vacuum cleaner is not intended for use in a humid environment, this must be included in the user's manual. This and many other risks can be indicated by a producer who has carried out the standardized risk assessment. The producer writes the user's manual for the vacuum cleaner, including the safety precautions, on the basis of the risk assessment.

Question 7
I have bought a metal-cutting machine for use in France, but it came with a user's manual in German. Can I accept it as is?

The producer or importer who sells products which fall under the scope of the Machinery Directive on the French market is responsible for the French user's manual. Your direct supplier is responsible for correct delivery of the product including a correct user's manual. If your supplier is a German producer, he or she should provide a French user's manual. If you have bought your metal-cutting machine from a French importer, the latter should remind the German producer (in writing) of the fact that a user's manual in French should be supplied.

Question 8
Our company has bought a packaging machine for our own use which has a CE Marking. I understand that the designation "CE Marking" indicates that this packaging machine must be safe. In the light of my extensive practical experience, I have doubts about the safety level of the machine. Even in the user's manual, only cursory attention is paid to safety. What can or must I do?

The producer or importer of your packaging machine within the EEA is responsible for affixing the CE Marking. This producer/importer is responsible and liable for the safe use of the product for at least ten years. If you have questions about the safety of your packaging machine and the correctness of the user's manual, you should first contact your supplier. In order to avoid misunderstandings, this contact should always be confirmed in writing (by registered letter). If you are not satisfied with the help you have received, you can engage an external expert. Inform your supplier accordingly and in writing about the further state of affairs.

Question 9
For over twenty years our company in Liverpool has been buying lawn mowers from Italy. For practical reasons our company has translated the user's manuals into English. Can we continue to do this without a risk assessment?

The Italian producer is responsible for supplying a safe product and a correct user's manual in English. If the Italian producer does not supply an English user's manual, you yourself must bear the responsibility for a correct user's manual. If problems arise as a result of your translation, your company is jointly responsible and liable for any imperfections. Assuming your company decides to continue with translation, you can always submit your translation for approval to your Italian producer.

The basis of a correct user's manual is formed by a legally compulsory risk assessment. If the risk assessment has already been carried out by the Italian producer, you can limit yourself to the translation of the Italian user's manual.

Question 10
Our company must compile a Technical Construction File prior to selling circular saw machines. Is the execution of a risk assessment sufficient?

The risk assessment forms a major part of the Technical Construction File. Other parts of this file relate, among other things, to technical drawings, calculations, test reports, and a copy of the user's manual and the EC Declaration of Conformity.

A circular saw is on the list of dangerous machines (Annex IV to the Machinery Directive). This means that before sale you must submit your circular saw machine to a Notified Body. This official body then inspects both your circular saw machine and the associated Technical Construction File according to the applicable (European) standards. Only when a Notified Body has given its approval can you launch your circular saw machine on the market within the EEA.

Question 11
We sell a simple stapling machine. Can we exclude responsibility and liability for accidents by including safety precautions in the user's manual?

Certain stapling machines fall under the sphere of operation of the Machinery Directive and they mostly seem to be simple products. An important formality arising from this Directive is to carry out a risk assessment. Only the result of a (standardized) risk assessment as described in this book will give you a definitive answer on the applicable safety measures. There is a hierarchy in taking the correct safety measures. First of all a definitive technical solution must be sought. If technology does not provide a solution, the protection should be installed. If this too is not a realistic option, only then can you fall back on the use of good safety precautions.

Note: A bench model of a stapling machine does not fall under the sphere of operation of the Machinery Directive. The spring in this machine is not the energy source but is used to advance the staples. A staple gun, on the other hand, is indeed a machine because the spring is the energy source.

Question 12
How long must the producer within the EEA keep the user's manual and EC Declaration of Conformity in the Technical Construction File?

The producer is responsible and liable for the safe use of a product for a period of at least ten years. As the user's manual forms an essential part of the information supply, this document should always be kept with the machine. The EC Declaration of Conformity must also be kept for at least ten years.

Question 13
Our company in the Netherlands bought a winding machine in Switzerland for our own use which has a CE Marking. The Swiss producer supplies a user's manual in German. Must we, as the purchaser of the winding machine, translate this user's manual?

Switzerland is not one of the countries within the European Union, nor is it part of the European Economic Area. Switzerland must therefore be regarded as a "third-party country" as regards the EU and EEA. This means that the Swiss producer must not affix the CE Marking to the winding machines. The importer who brings the winding machine from Switzerland onto the European market is responsible for the safety of this machine. Your company should therefore perform a certification procedure for the CE Marking and supply the obligatory user's manual.

A winding machine falls under both the Machinery Directive and the EMC and Low-Voltage Directives. It is best for you to inform your Swiss supplier of this. It is possible that your supplier has gained more experience of the CE Marking with other users in the EEA (i.e., outside Switzerland). Your company may be able to make use of the efforts made by other European users of the winding machine. If this is not the case, then you must move onto self-certification or outsource this. A brief summary of your obligations: compiling a Technical Construction File, writing user's information in the language of your country, and drawing up and signing the EC Declaration of Conformity. The Technical Construction File will certainly have to include a risk assessment.

Fifteen countries belong to the European Union (EU):

France	Sweden	Finland
The Netherlands	Belgium	Luxembourg
United Kingdom	Ireland	Germany
Portugal	Spain	Italy
Denmark	Greece	Austria

The EEA is made up of the fifteeter of these machines is responsible and liable for the ability to use these machines safely for a period of at least ten years. The Directive which is applicable to earth-moving machines is the Machinery Directive.

Question 14
What requirements are place on used earth-moving machines from the former Soviet Union?

Used earth-moving machines from outside the EEA must comply with the applicable European Directives and be give the CE Marking. The importer of these machinese is responsible and liable for the ability to use these machinese safely for a period of at least ten years. The Directive which is applicable to earth-moving machinese is the Machinery Directive.

Question 15
We supply circular saws to a customer in Belgium. In which language must we write the user's information?

Your company can simply check in which language area your customer is based. This language area within Belgium determines the language of your user's information. If you want to be absolutely correct in all cases, you can write your user's manual in both Dutch and French. The reason for translating the user's manual is based on the fact that you want to be certain that the user understands the instructions for safe use, maintenance, and repair.

Question 16
We are an American company that exports laser machines to Europe. We do not include a risk assessment. Is it still possible to sell a laser machine in an EEA country without this document?

The importer within the EEA is responsible for a safe laser machine. A laser machine which is launched on the market in Europe must comply with the requirements of the Machinery Directive, the EMC Directive, and the Low-Voltage Directive. If the laser machine has been manufactured according to these Directives, the CE Marking can be affixed. The European producer or importer declares by means of the CE Marking that the machine can be used safely. Although you, the American producer, are not obliged to do so, you can supply a risk assessment in accordance with the requirements of the CE Marking. In other words, you may demonstrate that the laser machine complies with the European requirements. However, the importer should verify this and be able to confirm it. For this he or she needs all the information of the Technical Construction File. The risk assessment is an important part of the Technical Construction File.

Question 17
An inspection by the European Labor Inspectorate established that there was no Technical Construction File for a cutting machine we produced ourselves. Using a risk assessment it was demonstrated that the machine does not comply with the safety requirements. What is a Technical Construction File? How can I compile one, and who can ensure the safety of my cutting machine?

The producer or the importer of a cutting machine within the EEA is responsible for compiling a Technical Construction File. A Technical Construction File consists of all the technical drawings and calculations which are used to gain an insight into the safety of the machine. For a cutting machine, a risk assessment as described in this book is also compulsory. If standards are used in the design or production, the Technical Construction File must indicate which ones.

If risks are established during the risk assessment which are unacceptable, these should be solved both technically and by means of information. This means not only that the dangers are pointed out, but also that the design is modified or protective measures are taken.

Question 18
Our directors have decided to build a new factory for the baking of potato chips. As the person responsible for this, I am busy approaching various suppliers and one installer. Upon inquiry, one supplier stated that they deliver with a IIA declaration and another that they deliver with a IIB declaration. What is the difference between these two declarations and which delivery is the most acceptable?

The difference between a IIA and a IIB declaration can be shown as follows:

The *IIA declaration* is an EC Declaration of Conformity and is given only if a machine or installation complies with the Machinery Directive. The producer or importer within the EEA submits this IIA declaration and states on it, among other things, the applicable Directives and standards, along with his or her company name and address. This EC Declaration of Conformity can be seen as a feature of a machine or installation. The purchaser or user will have to keep this IIA declaration with the machine or installation for a period of at least ten years.

The *IIB declaration* is a manufacturer's declaration. A producer or importer within the EEA of a part or semi-finished product declares only that a part or semi-finished product has been supplied. In other words, this covers parts and semi-finished products to which the CE Marking is not directly applicable. In the event of damage, the supplier concerned cannot be held responsible for the safety of the entire machine or installation.

Question 19
Is a risk inventory the same as a risk assessment?

The risk assessment applies to products which fall under the Machinery Directive. A risk inventory (and evaluation) is part of the working conditions legislation. There are similarities between the two, but a risk inventory and evaluation (also called RI&E) is more global in structure than the risk assessment as described in this book. The risk assessment which is carried out within the framework of the Machinery Directive is generally done in the design phase by the producer. A risk assessment goes deeper into the risks of the product itself. An empolyer is obliged within the framework of the Law on Working Conditions to make a risk inventory and evaluation of all the dangers present in the company. Thus, a risk inventory and evaluation of work equipment is covered by this. The new working conditions rregulations refer to EN 1050 relating to work equipment. This standard is not a legal requirement but a guide for achieving the prescribed level of safety.

Question 20
Can I simply use the user's manual of a colleague (competitor)?

It goes without saying that the use of another person's material without permission is not allowed. Most user's manuals are protected by copyright. Material may not be used without the permission of its legal owner. Furthermore, the extent to which the manual of a "colleague" actually applies to your product is questionable. This is only possible if the products are completely identical. It is clear that the basis of a good user's manual is the risk assessment. Only by carrying out a risk assessment can you determine exactly what information must be passed on to the user. The user's manual then even gives the advantage that it provides you with legal arguments in the event of damage through misuse of the product.

Question 21
A customer is asking for the Technical Construction File for a brick machine supplied by us. Do we have to submit the entire file, or will just a part suffice?

The Technical Construction File does not need to be supplied to your customers. They receive a correct user's manual and an EC Declaration of Conformity from you. These documents form part of your Technical Construction File, so in that sense you have indeed supplied part of the Technical Construction File.

Only European inspection bodies such as the Labor Inspectorate and the Inspectorate of Goods are entitled to inspect the Technical Construction File. It depends greatly on the situation, but you could be open with part of your file. It is advisable to consult an expert in advance on the background to this request.

Question 22
Does our milling head fall into the category of interchangeable equipment under the Machinery Directive? Must I also write a user's manual for a piece of interchangeable equipment?

A piece of interchangeable equipment is a part which can change the function of a machine or installation.

Example: if your milling head can be installed on a universal drill column, this means that your milling head changes a drilling machine into a milling machine. By using your milling head, a new machine is formed with different risks. The producer of a drill column can never be held responsible for damage (injury) which arises from the use of this milling head.

A piece of interchangeable equipment must certainly comply with the Machinery Directive 89/392/EEC, and therefore a risk assessment must be carried out and the milling head supplied with a user's manual.

Question 23
As the hire company for machines, must we ourselves carry out the risk assessment for our line? Can we give the original user's manuals to our customers?

A hire company for machines can limit itself to providing the information from the producer. Assuming that your hire company only hires items to third parties, it is not necessary to carry out a risk assessment. However, at the same time it is necessary for you to check whether the requirements placed on the product are still met before the product is hired out again. For example, check that all the protective covers and safety measures are still present. This check is best integrated into the maintenance check on return of the product. Furthermore, the risk assessment can play a role in this.

It is also compulsory for the user's manual to be supplied with the product at all times. Hirers are often laypersons and should be clearly instructed on what is and is not permitted. You could even legally specify the delivery of the user's manual.

4.3 PRACTICAL EXAMPLE OF RISK ASSESSMENT

4.3.1 Introduction

As described in Section 3.2.3, risk analysis consists of five steps: risk identification, risk assessment, risk evaluation, risk reduction, and safety evaluation. This section explains the risk assessment for the Machinery Directive by means of a practical example.

The American company Hedge Inc. designs and manufactures garden tools, including hedge clippers. Because Hedge Inc. sells the hedge clippers on the Dutch market, this company has to deal with, among other things, the CE Marking for the Machinery Directive.

4.3.2 Step 1: risk identification

As part of risk identification, it must be clear:

- What product is concerned
- For whom the product is intended
- Under what conditions the product will be used

4.3.2.1 Identification of the machine. Figure 4.1 shows the identification data of the machine such as the manufacturer, the product name and model, and the description.

4.3.2.2 Identification of users. Hedge Inc. sells the hedge clippers on the Dutch consumer market. Users can be identified as follows:

Manufacturer	Hedge Inc.
Product	Hedge clippers HSW-01
Description	Machine for clipping hedges
Clipping diameter	Maximum 14 mm
Blade	Cuts on both sides, 500 mm cutting length, 1700 cutting movements per minute
Weight	3.0 kg
Power supply	220–240 V/50 Hz/1.8 A
Power consumption	1400 W

FIGURE 4.1 Identification data of the hedge clippers

Operator: Dutch inhabitants
Repairer: Recognized professional

In principle, every Dutch inhabitant, irrespective of gender, training, or age, should be able to use the hedge clippers. An exception is children: they must not operate the hedge clippers.

The average consumer must not be expected to be able or permitted to perform repairs to the hedge clippers. Repairs should be left to a recognized professional.

4.3.2.3 Identification of conditions of use. "Conditions of use" refers not just to normal use; use which can reasonably be expected also plays an essential part. The Machinery Directive prescribes this in law in Annex I, 1.1.2.c:

> *During the design and construction of the machine as well as when writing the user's instructions, the manufacturer must not just assume normal use of the machine but also use which can reasonably be expected.*

An example of use which can reasonably be expected is that a consumer might use the hedge clippers as a universal saw.

Under Annex I, 1.1.2.a, of the Machinery Directive, account should also be taken of foreseeable abnormal conditions of use:

> *The (safety) measures taken must be geared to eliminating completely any risk of accident during the expected service life of the machine, including assembly and dismantling, even if these risks are the result of foreseeable abnormal conditions.*

For example, it is certainly not inconceivable for the consumer to be caught in a rainstorm during trimming and deciding to "just keep on working." Or another example: The consumer has a go at a snow-covered hedge with the hedge clippers.

A major assumption for the manufacturer is the expected service life. In other words, the manufacturer should take account of the safety of the machine throughout its service life, with a focus on parts which are subject to wear and deserve particular attention.

The following conditions of use are reasonable to assume for the hedge clippers:

- Exterior use
- Use with an extension cord
- Use in the rain and in wet and humid conditions
- Use in the garden
- Inappropriate use
- Use in another country than where purchased
- Use in a defective or damaged state

However, the identification of all conditions of use can never be exhaustive. Briefly, the manufacturer must be aware of the following important aspects:

- Normal use
- Abnormal use

- Use which can reasonably be expected
- Safety requirements which apply for the entire service life of the machine

The examinations of risks, the user, and the conditions of use of the machine are always included in detail in the risk analysis.

The identifications carried out form the frame of reference within which the examination takes place. A change in one of the variables has consequences for the safety risks. For example, suppose Hedge Inc. decides to concentrate on the professional market and no longer on the consumer market. There will then be a different target user group.

4.3.3 Relationship with the Machinery Directive

Are hedge clippers a machine?. Before the procedure for compliance with the requirements of the Machinery Directive begins, it important to ask whether the product concerned actually falls under the area of application of the Machinery Directive. This can easily be tested by examining whether the machine matches the definition of a machine as cited in Article 1, Section 2.

Hedge clippers are a machine because:

- They are a composition of mutually connected parts.
- There is a drive mechanism (electric motor).
- The hedge clippers have a specific application.

Are hedge clippers an exception? It is possible for a product to be a machine according to the definition, but also to be one of the exceptions to the Machinery Directive. It is therefore always sensible to read through the exceptions (see Article 1, Section 3). However, the electric hedge clippers do not fall under these exceptions. Manual hedge clippers, on the other hand, operate on the physical energy of a person and are not classified as a machine.

Involve Notified Body? If a machine is included in Annex IV of the Machinery Directive, a Notified Body must be involved in the CE Marking procedure. However, the hedge clippers are not in Annex IV, and it is therefore not necessary to involve a Notified Body.

4.3.4 Specific requirements and standards

4.3.4.1 Specific requirements. Annex I of the Machinery Directive states the requirements which apply for machines. The first chapter of this Annex names the general requirements which are applicable to every machine. The specific requirements can be found in Chapters 2 to 6 of Annex I.

Annex I, 2.2, gives the specific requirements placed on portable machines held and/or controlled by hand. The following specific requirements are applicable to the hedge clippers:

Portable machines which are held or controlled by hand by the operator must comply with the following basic health and safety requirements:

- Depending on the type of machine, the machine must have an adequate supporting surface and a sufficient number of provisions for gripping and holding the machine with the correct dimensions and applied in the correct positions so as to ensure the stability of the machine in the operating conditions specified by the manufacturer.
- Unless it is technically impossible or if there is an independent operating unit, if the hand grips cannot be released in the event of danger, the machine must have operating units for starting and/or stopping it which are positioned in such a way that operation is possible without the operator releasing the hand grip.
- The machine must be designed, built or equipped in such a way that there is no danger if it is temporarily started and/or remains in operation after the operator has released the hand grips. If this regulation cannot be executed technically, compensating provisions must be made.
- A portable machine held by hand must be designed and built in such a way that, if necessary, contact of the tool with the processed material can be inspected visually.

User's instructions

The user's instructions must contain the following information on the vibrations produced by machines held and controlled by hand:

- The weighted root mean square value of the acceleration to which the arms are exposed if this acceleration is greater than 2.5 m/s^2, defined according to the appropriate test regulations. If this acceleration is not greater than 2.5 m/s², this must be stated. If there are no applicable test regulations, the manufacturer must indicate by what measuring methods and under what conditions the measurements were taken.

With reference to the second requirement, it must be recognized with the hedge clippers that it is dangerous to operate the machine with one hand. First, the operator then has too little control over the hedge clippers. Second, the operator may come into contact with the blade with one hand. The hedge clippers are therefore required to be designed so that only two-handed operation is possible.

In the third requirement, particular attention is paid to the operating units. For example, it is possible to interpret from this that when the clippers are put down in any state whatsoever, it must not be possible for them to be started up by their own weight.

4.3.4.2 Specific standards. The application of (harmonized) standards is not obligatory. Nevertheless, there are good reasons for using the standards, among other things because:

- It leads to a "Presumption of Conformity."
- It forms a concrete technical guide.

The following standards are applicable to the hedge clippers:

- All relevant type A standards
- EN 294 "Safety of machines: safety distances to prevent upper limbs from reaching hazardous zones"
- EN 574 draft "Safety of machine: two-handed operation"
- Ergonomics standards
- Vibration standards

Other Community Directives, such as the EMC and Low-Voltage Directives, are also applicable to the hedge clippers. The hedge clippers must also comply with these Directives before the CE Marking can be affixed.

Step 1: Manufacture	Production Packaging
Step 2: Transport	Transporting to wholesaler or retailer Transporting from retailer to customer
Step 3: Preparing for operation	Unpacking Removing packaging
Step 4: Starting up	Connecting extension cord Plugging unit into socket Moving to the trimming location
Step 5: Use	Switching on hedge clippers Making trimming movements Walking with the hedge clippers Inappropriate use
Step 6: Finishing use	Switching off hedge clippers Disconnecting electric cable Storing
Step 7: Maintenance	Cleaning Lubrication
Step 8: Repair	Replacing defective parts
Step 9: Disposal of the machine	Dismantling Waste disposal or recycling

FIGURE 4.2 Product process phases for the hedge clippers

4.3.5 Step 2: Risk assessment

4.3.5.1 Product process phases and their purpose. As an aid for setting up the risk assessment for the hedge clippers, the life cycle of the hedge clippers is divided into a number of successive steps, the so-called product process phases. The aim of this method is to provide a structure which forms the basis for the risk assessment. If the product process phases are followed step by step, there is a smaller chance of specific risks being overlooked.

4.3.5.2 Description of product process phases for the hedge clippers.
Figure 4.2 describes the product process phases for the hedge clippers. Using the product process phases, the risk assessment can then be drawn up.

4.3.6 Step 3: risk evaluation

The analysis of the hedge clippers highlights various risks which may arise during use. The major risk categories are:

Category	Description
A	Trapping or injury of fingers between the blades
B	Catching of hair or loose clothing in the blades
C	Cutting through the electric cable during trimming
D	Use in a humid environment (risk of short circuit)
E	Other inappropriate use

4.3.7 Step 4: risk reduction

If all the risks are known, measures can be taken to reduce the risks to an acceptable level. The following sequence of priorities is the starting point:

- Intrinsically safe design
- Taking protective measures
- Providing information to the user

4.3.7.1 Intrinsically safe design. With the hedge clippers the greatest danger is in the moving part: the blade. The movement of the blades is also the heart of the machine's function. Therefore it is not possible to screen the moving parts in such a way that the risk is reduced to zero. However, it is possible to adapt the shape of the blades in such a way that the risk of injury is minimal—for example, so that when using reclining blades, objects with a diameter greater than 14 mm cannot be cut.

4.3.7.2 Taking protective measures. The following two measures reduce the danger from the moving parts:

Two-handed operation. By incorporating an extra switch in the hand grip or a bracket on the hedge clippers, the user is required to hold the machine with both hands to be able to start the machine. The risk of the user's injuring his or her fingers is thereby minimized. **Protective sheet.** A protective sheet of Plexiglas between the hand grips and the blades gives extra protection:
- The sheet prevents the operator's hand from touching blades which are still moving if the hand unexpectedly slips from the hand grip or bracket.
- The sheet also offers protection against trimmed debris which might fly back.

The following two measures can reduce the electrical risks:

- Double insulation
- Power cord with a circuit breaker and grounded plug

4.3.7.3 Providing information to the user. Despite protective measures, the hedge clippers remain a product which always entails risks during use. Therefore it is important to give the user sufficient information about these risks. The following is a suggested set of safety precautions that might be provided for the hedge clippers on the basis of the identified risks both in a general sense and using risk categories.

General
- Although as many measures as possible have been taken to guarantee the safety of the hedge clippers, working with hedge clippers is and remains a dangerous activity. Look out for yourself and your environment at all times!
- Remove the plug from the socket immediately after using the hedge clippers.

- Never leave the machine unattended while it is still connected to the power.
- Never leave an extension cord unattended while it still has a voltage running though it.
- Remove the extension cord from the socket before working on or with the hedge clippers for anything other than actual cutting.
- Do not operate in the vicinity of other people or animals.

Category A: Trapping or injury of fingers between the blades

- Always keep hands and other body parts away from the vicinity of the cutting blades.
- During operation always hold the hedge clippers firmly on the designated hand grips: this is the safest method of operation. Only operate the hedge clippers using two hands on the hand grips.

Category B: Loose clothing or hair in the blades

- Wear clothing which lies close to the body. Avoid scarves and other loose clothing.
- Take measures if you have long hair so that it cannot come into contact with the machine.

Category C: Cutting through the electric cable during trimming

- Keep the power cord and extension cord away from the blades. Feed the cable from behind your back.
- Remove the plug from the socket immediately before doing anything else if the extension cord is accidentally cut.

Category D: Use in a humid environment (risk of short circuit)

- Only cut when the plants are dry and certainly not when it is raining.
- Store the hedge clippers indoors after use. Never expose the device to rain.

Category E: Other inappropriate use

- Keep children away from the machine when it is plugged in.
- Do not allow children to operate the hedge clippers.
- Only use the hedge clippers for cutting plants and thin branches of bushes and hedges. It is inadvisable and impermissible to cut other material.

4.3.8 Step 5: safety evaluation

Once the risks that might occur during use of the hedge clippers have been reduced, an evaluation follows. The question is whether the risks have been reduced to an acceptable level. This is now examined for each risk category.

Category A: Trapping or injury of fingers between the blades

This risk is sufficiently reduced by two-handed operation in combination with the protective sheet.

Category B: Loose clothing or hair in the blades

This risk has been reduced by using a protective cover in combination with safety precautions. The risk is not completely eliminated, but is reduced sufficiently.

Category C: Cutting through the electric cable during trimming

Double insulation of the electric cable in combination with safety precautions reduces the risk. The risk of occurrence is not reduced very greatly but the risk of injury is.

Category D: Use in a humid environment (risk of short circuit)

Double insulation of the electric cable in combination with safety precautions reduces the risk. The risk of occurrence is not reduced very greatly but the risk of injury is.

EC DECLARATION OF CONFORMITY FOR MACHINES

(Directive 89/392/EEC, Annex II, under A)

We,

Manufacturer: Hedge Inc.
Address: 498 Bush Square, Bethesda, Maryland 20810

Do hereby declare that

ELECTRICAL HEDGE TRIMMER HSW-01

Is in conformity with the essential requirements of the Machinery Directive (Directive 89/392/EEC, as recently amended), especially the relevant requirements for this machine of Annex II of the Machinery Directive.

Level of the sound pressure in accordance with 86/188/EEC and 89/392/EEC, measured by DIN 45635: L_p (sound pressure) < 86 dB (A).

The weighted quadratic average value of the acceleration in accordance with EN 50144: < 2.5 m/s^2.

Is in conformity with the following insulation-requirements:
The machine is insulated by VDE 0740 class II and CEE 20 and therefore can be connected to sockets without grounding.

Drawn up in: Bethesda, Maryland
On: 19 December 1997

Signature:

By: Hedge Inc., R.J. Hedgeman

$C\epsilon$

FIGURE 4.4 The EC Declaration of Conformity

Category E: Other inappropriate use

Inappropriate use of the hedge clippers is always possible even when this has been warned against. Therefore it is important to reduce the risk of injury by means of all the safety measures together.

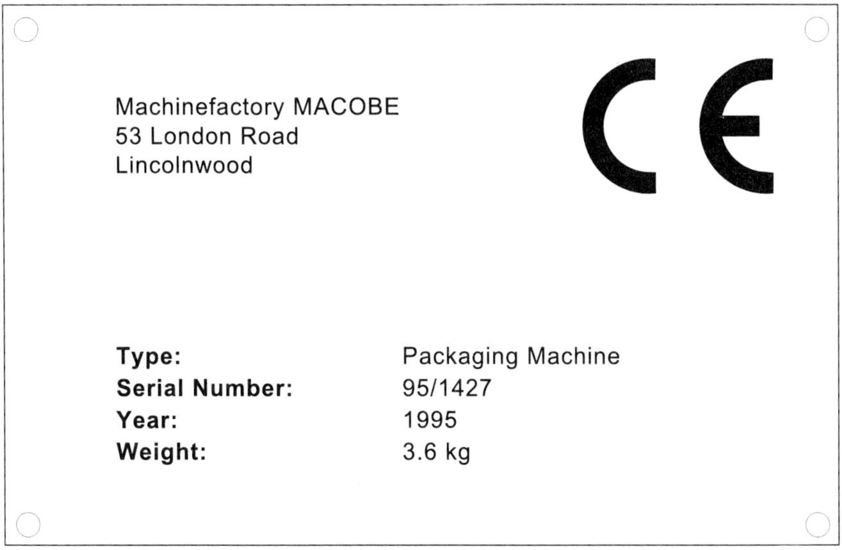

```
┌─────────────────────────────────────────────┐
│                                             │
│   Machinefactory MACOBE          C €        │
│   53 London Road                            │
│   Lincolnwood                               │
│                                             │
│                                             │
│   Type:              Packaging Machine      │
│   Serial Number:     95/1427                │
│   Year:              1995                   │
│   Weight:            3.6 kg                 │
│                                             │
└─────────────────────────────────────────────┘
```

FIGURE 4.5 Example of the CE Marking of conformity for a packaging machine

General conclusion. By concluding the five-step plan, Hedge Inc. sees to it that no major risks associated with the use of the hedge clippers have been ignored.

4.4 EC DECLARATION OF CONFORMITY

Finally an EC Declaration of Conformity is drawn up for the hedge clippers as described in Chapter 3. See Figure 4.4.

4.5 AFFIXING THE CE MARKING

The typography of the CE Marking should meet the requirements as set in Annex III of the Machinery Directive. If the dimensions of the machine allow it, the letters or characters used should have a minimum height of

5 mm. A deviation from this is permitted on smaller machines, but the ratio of the CE initials must not change. See figure 4.5 for example.

The CE Marking must be affixed to the machine in a permanent manner, for example, with an engraved aluminum plate attached to the machine with rivets.

The following should be included on the plate with the CE Marking:

- The initials CE
- The full name and address of the manufacturer or an authorized representative
- Type or name of machine
- Serial number if applicable
- Year of construction
- Total weight

5
Liability within the EEA

5.1 INTRODUCTION

The reason there is a chapter on liability in a book on the obligations arising from the Machinery Directive is that these obligations can have a direct or indirect relationship to accidents with products and the consequent question of liability. User information and user manuals, or the lack of these, have a direct influence on whether a product is adequate. Thus, a manufacturer can be held liable for damage caused by inadequate instructions. Of course, a manufacturer can also be held liable for damage caused by inadequate products in general. This chapter merely deals with the liability situtation of the producer and the importer within the EEA; however, the American producer or exporter will get a clear insight from it into the most important aspects of European product liability.

5.2 WHAT IS LIABILITY?

5.2.1 Definition

It can happen that the use of a product causes bodily harm to a person, or damage to the property of a person. This person can then hold someone—for example, the manufacturer of the product—responsible for the damage suffered and can seek compensation. In other words, the person can hold the manufacturer liable.

Liability may be defined as follows:

> *The legal duty to provide compensation for damages caused by a defect in a product.*

5.2.2 Who can be held liable?

In the case of a faulty product, it is the producer who is liable. In this connection, however, "producer" is not limited to the manufacturer. The manufacturer of a part of the product, an authorized agent, the importer who imports the product into the EEA, and the dealer who places the product on the market in the producer's name may all be considered producers.

5.2.3 When is someone liable?

If a user suffers bodily harm or dies as the result of the application or use of an end product, or there is damage to property, this can lead to a verdict of liability. In such a case, the damage must have been caused by a fault in the product. One speaks of a fault if the product does not provide the degree of safety that can be expected. There are three possible causes of a fault:

- Design errors
- Production errors
- Instruction errors

Having to correct *design errors* after production has started can have major consequences, because in such a case the entire production run may be faulty. *Production errors,* on the other hand, may affect only a single series of products with which something went wrong during production. What can be more serious is *instruction errors:* erroneous, unclear, and/or incomplete instructions for use. In such a case the producer can be held liable for keeping unsafe products on the market. If producers or dealers are aware that their product is unsafe, they have the duty to inform the public about the fault and, if necessary, do everything possible to withdraw the product from the market. In such a case one speaks of product recall liability.

5.2.4 Proving liability

If compensation is to be paid, a number of things will have to be proven. However, the burden of proof is not on the aggrieved party but on the producer! The aggrieved party must be able to show that he or she has suffered damage and that this damage was caused by the product in question. The guilt of the producer does not have to be proven. The producer must prove that there is no cause for being found guilty or held liable.

5.2.5 Exclusion of liability

In a number of cases it is possible to transfer liability. For the raw materials and parts that were used, the producer can shift liability to the supplier. The producer can also reach agreement with an intermediary to transfer the liability. If such an agreement includes a safeguard clause, then a producer who has had to compensate someone for damages can claim these damages from the intermediary. This may be possible even if the aggrieved party cannot hold the intermediary liable for the damage. In such a case the aggrieved party calls only the producer to account, after which the producer can recover the damages from the intermediary.

5.3 RELATIONSHIP BETWEEN CE MARKING AND PRODUCT LIABILITY

5.3.1 CE Marking—product liability

It can be argued that the CE rules and the Directive on product liability partially overlap and supplement each other: the CE Marking indicates that the product to which it is affixed complies with specific safety requirements imposed in the law. The affixing of the CE Marking is a legal obligation, and the Marking must (in principle) be affixed by the producer of the product. In other words, the producer may only circulate products which can be used safely by the user, and the product liability regulations state that the producer is liable for the product if it causes damage.

5.3.2 No complete exclusion of liability

Producers invest a great deal of money to produce and sell safe products in accordance with the legal requirements. Compliance with these legal safety requirements is reasonable proof of a good and safe product. The risk of damage being caused by that product therefore is probably small, but is not completely eliminated. It therefore remains possible for a producer of a product to comply with all the requirements from the New Approach Directives and nevertheless to launch a defective (unsafe) product in the sense of product liability. After all, the health and safety requirements imposed by these Directives give a minimum safety level with which the product must comply in any case. Product liability obliges the producer to pay compensation if damage is caused as a result of a lack of safety in the product. That remains the case even if the legislation and standards indicating a minimum level for safety are complied with.

Whoever has to certify the products is obliged to make the company or trading name and address known with the CE logo. This may be the producer, the authorized representative of the producer in the EEA, or the importer who imports the product into the EEA (if the producer is located outside the EEA). This obligation means that the responsible person can easily be traced and this therefore increases the risk of liability for the producer.

5.3.3 The advantage of CE Marking in relation to liability

However, it is not the case that the New Approach Directives only lead to an increased risk of liability for the producer. It is, of course, good for the producer to comply with the new rules imposed. Otherwise a "defect" is very simple to establish.

When producers fulfil their CE obligations, then the CE marking, in the light of legal liability, can also work to their advantage. This is brought about by the nature of the obligations.

Producers must not only themselves adapt the product to the law, but they will also have to submit a technical file, supply a correct user's manual, and sign a Declaration of Conformity. In this way they have put in black-and-white their position as regards the risks of the product and how they have reduced unacceptable risks. For example, the responsibility for parts and their use can also be better separated and transferred. If it can be demonstrated that the producer has made sufficient efforts to comply with the law, any amount which may have to be paid out will be considerably reduced. Parts of the CE obligations, such as the Technical File, the user's manual, product documentation, and the Declaration of Conformity, play an important role in all this as evidential material. For further information on the legal power of proof of the user's manual, refer to Section 5.5.

5.3.4 Establishing contractual matters

The CE Marking does not rule out product liability, but it can limit the risk of liability. The occurrence of a financial risk after a case of damage can also not be entirely ruled out. Whether a producer will ultimately have to bear the damage will depend, among other things, on the contractual relations with suppliers and purchasers. CE Marking does not exclude contractual liability, in contrast to legal liability.

The producer must ensure:

- Good purchasing conditions (or purchasing contracts) to suppliers of parts and semi-finished products
- Good sales conditions in transactions with purchasers

5.3.5 Insurance

Insurance against liability is possible. There is both general liability insurance (for companies and professions) and product liability insurance.

A misunderstanding with this type of insurance is that the compensation amount will always be paid. At the heart of professional liability insurance is liability for damage suffered by third parties. The policy contains restrictions on the damage to be compensated; the various exclusions in the policy also have a function of limiting cover. The liability for a product is generally also insured under the company insurance policy conditions. However, transaction damage is excluded in all cases—in other words, damage as a result of replacement, improvement, or repair of *supplied products* as well as damage resulting from inability or insufficient ability to use the products.

This type of policy also excludes so-called negative default. This means that damage as a result of nonobservance or late observance of an agreement as well as the damage arising from a penalty, compensation, guarantee, indemnification, or other similar condition is not compensated. Nonobservance of an agreement can mean the production and sale of a product which does not (completely) comply with health and safety requirements and consequently does not have the CE Marking or has no bona fide CE marking. It may therefore be that payment will not be made if the CE Marking has not been affixed or has not been affixed in good faith if it should have been.

5.3.6 Criminal liability

Besides the civil product liability cited above, there is also criminal liability for products. If products do not or do not fully comply with the legal requirements, criminal sanctions can be imposed. The imposition of such sanctions can take place both after a case of damage and after a preventive check.

If it is established after a case of damage that the product does not comply with the legal requirements imposed, criminal sanctions can follow for the person who is responsible for compliance with these requirements. The aforementioned civil product liability also remains in force. This criminal liability exists for the producer even if the product has not (yet) caused any damage. Inspection bodies can inspect the goods put into circulation for compliance with the imposed requirements.

5.4 PREVENTIVE MEASURES: STEP PLAN

A producer will want to limit damage as far as possible or preferably eliminate it before any case of damage takes place. A step plan is included below which indicates by means of points how a producer—both of an end product and of raw materials, semi-finished products, or part of a product—can limit or eliminate risks of liability. Basically this step plan is meant for the European producer or importer within the EEA; however, this list may also be of help for the American exporter or producer who does business with European countries.

1. Design and production phase

Producers should take account of the applicable safety requirements and standards relating to their products from the first design. They must design and produce safe products. The safety aspects of a product must also be taken into account in a marketing plan. The product developments and specifications (regulations, standards, state of the art, etc.)

should be documented both for internal policy and for evidence in the event of any case of damage.

Besides the increasingly strict safety requirements, the (safety) expectations of the users and possible uses are also subject to change. Account must likewise be taken of the safety implications of user's instructions, marketing communications, packaging, warnings, etc. This should all be constantly examined from the point of view of the user, not forgetting the careless user. In other words, use which is not (completely) normally expected must be taken into account in design and production.

2. Safety and quality control procedure

It is advisable to draw up procedures for safety and quality control during the entire production process. For this the producer would have to introduce integral quality assurance, such as the ISO system, into the company. If the products are particularly dangerous it is advisable to have the purchasers carry out additional safety inspections.

3. Procedure for removing unsafe products from the product line

It is advisable to draw up procedures for removing unsafe products from the product line and from the production process. In order to discover whether a product is unsafe, the producer should have the product tested throughout the entire production process.

4. Procedure for handling complaints

It is advisable to draw up procedures for systematically handling complaints. The following points are important:

- Easy accessibility of the producer's company by the complainant or victim
- Instructions for the receiving company
- The internal organisation of the receiving company
- The settlement to the complainant

5. Procedure for removing unsafe products from the market

It is advisable to draw up procedures for product recall. Consideration must not only be given to the actual recall of unsafe products from the market, but also to informing those who use the product.

6. Supply of information

The producer shall have to ensure effective and efficient supply of information. Consideration must not only be given to the external supply of information, but also to the internal supply of information. Everyone involved within the producer's own company must be well informed, and a system must be set up which makes communication more flexible. Because the producer is the first link in the production and distribution chain, the company will have to ensure that the correct information reaches purchasers and supplier. The information coming from outside the company must also be properly included and used.

7. Information for employees

Any producer that is also an employer should inform employees within the company on safety and liability. In this way the employees can take account of the producer's responsibilities in their work. The employees must be motivated to produce safe products and to prevent product liability claims in the most direct way.

8. State of science and technology

The producer must take account at all times of the constantly changing state of science and technology. The most up-to-date knowledge must be drawn on in all activities, in particular the incorporation of safety features, within the company.

9. Inspection of products supplied

On purchase, the safety of incoming raw materials, semi-finished products, or parts must always be checked. In the first place suppliers of raw materials, semi-finished products, and parts should be selected carefully. Suppliers with quality certification are preferred, primarily in the case of imports from outside the EEA. Extra consideration must be given to the safety (and quality) of the raw materials, semi-finished products, or parts because legal requirements and standards vary from country to country. In such cases ask for additional guarantees and assurance from suppliers from non-EEA countries.

The producer should avoid buying raw materials, semi-finished products, and parts from unknowns, because of later options of recourse in the event of damage. A producer who is held liable for paying for damage suffered which is caused by a specific raw material, semi-finished product, or part will want to recover this financial risk from the producer of the raw material, semi-finished product, or part. If the identity of this producer is unknown, then whoever has been held liable in the first place must pay the entire amount.

Also make a good examination of producers themselves for their conditions and/or contracts with or from suppliers because of exoneration conditions and/or indemnification conditions. You should be aware of what has been agreed and who is liable.

10. Written establishment of procedures
Always put all the above procedures and other activities in writing. This will lead to a flexible and clear course of affairs within the company. Many written documents will also be able to serve as evidence in the event of legal proceedings. Among other things, it is important to put into writing and to document the date of the launch of products, the inspections made, and product specifications.

11. Adapting contracts
The producer should always be aware of the contents and significance of general conditions and contracts in transactions. Existing conditions and contracts must be adapted regularly to new legislation. Producers of end products, raw materials, semi-finished products, and parts must consider the amendments to the New Approach Directives and other applicable laws. Especially for the American producer or exporter who does business with European countries it is important to have good contractual agreements with the importer in Europe. Importers within the EEA are responsible for the CE Marking and the products; however, they might want to settle the terms of their responsibility in a specific manner and have them put into a contract.

12. Safety and quality
It is advisable to use safety and quality measures in the entire marketing approach of a company, first to launch safe, high-quality products on the market as a company, and second, so that the producer can be distinguished from competitors through superior quality and safety.

5.5 APPROACH IN THE EVENT OF DAMAGE AND CLAIMS: CHECK LIST FOR THE END PRODUCER

A producer will do much to produce a safe, high-quality product. Nevertheless, everything can still go wrong after the product is launched on the market. Does the product still display a defect despite the certification with the CE-marking? Has the user used the product for a purpose other than that for which it was intended?

Below is a check list showing point by point what a producer must do and must think about in the event of a disaster. It also indicates what the producer must do to limit further damage. This check list is for both the end producer and the producer of a raw material, semi-finished product, or part. The latter category of producer in particular is not obliged to produce in accordance with the New Approach Directive, but does have to deal with product liability. An end producer is taken to mean the producer of complete products which may be intended for both the consumer market and business.

1. Information

If the victim makes a claim against the producer, the latter should contact a number of people or bodies, in the first place, an expert legal adviser (inside or outside the company) to protect the interests of the producer at a legal level. The insurance company with which the producer has taken out product liability insurance should also be informed about the claim made for damages. The producer will want the amount of compensation that has to be paid to be reimbursed by the insurance company. Not least, the person responsible for safety and quality control must be informed and engaged for the further proper settlement of the matter, not only as regards the victim, but also for the company itself.

2. Product

Examining the product that caused the damage can help indicate whether the product is the producer's own. The CE Marking can give information that helps a victim identify the producer of a faulty product; producers can make use of the same information in their defense.

3. Production and sale

If the product has been produced by the producer held liable, it should be discovered where and when the product was produced, among other things, in order to discover from which batch, if applicable, the product originates. It should also be discovered when and to whom the product (or batch of products) was sold. If the entire batch has to be recalled, it should be clear with whom the products are in use or in stock.

4. Launch on the market

Producers should also investigate whether faulty products in liability cases were actually launched on the market by them. If they find that this is not the case, then they can use this information as a defense in any product liability proceedings.

5. Gathering information

The producer must gather information about the victim. It is important to know whether it is a consumer who suffered the damage or an unfortunate passer-by (third party). If it is a company which is concerned, it is important to know whether it is a wholesaler who sold the product or a company which made modifications to the product.

6. Defective

The producer must find out whether the product was defective, i.e., unsafe, in relation to the degree of safety the user of the product might expect. The instructions and precautions, the judgment about what constitutes reasonable use, and the time when the product was launched on the market must be factors in making this determination.

7. Intended purpose

The specific purpose for which the user has used the product should be discovered. It is quite possible that the user has used the product for a purpose for which it was not intended. The intended purpose must be described accurately in the user's manual.

8. Time the defect arose

The producer should investigate when the defect arose. It is possible that the defect only arose later, i.e., after the product was launched on the market. If this is the case, the producer can use this as a means of defense in product liability proceedings.

9. Commercial or business purpose
The producer should examine whether the product:

- Was manufactured for resale or any other redistribution with economic purpose
- Was put into use or redistributed within the framework of conducting a profession or business

If neither is not the case, the producer may have a defense in product liability proceedings.

10. Compelling government regulations
If the product causing the damage was produced in accordance with compelling government regulations, the producer can use such compliance as a defense in product liability proceedings. The producer must be able to demonstrate which regulations these were and to prove that production was actually in accordance with these government regulations.

11. State of science and technology
The producer should examine whether it was possible to detect the defect at the time the product was launched on the market on the basis of the state of science and technology. If not, the producer can claim such limitations of technology as a defense.

12. Modifications to the product
It is possible that a safer product has been launched on the market after the product causing the damage was sold or that modifications have been made to the old product. Although a product may not be regarded as defective simply because a better product was launched on the market at a later date, the producer is obliged to send out warnings to those affected to inform them of the existence of a safer product and about any unsafe features of the old product.

13. Date of launch on the market

The precise date of launch on the market should be examined. If this date is before the enforcement dates of the applicable Directives and laws, these legislations are not applicable. The producer may then still be held liable under a different regulation, but these rules are less strict.

14. Own fault of the victim

It is possible that the damage may be the fault of the victim or of a person for whom the victim is liable. If this is the case, it may mean that the producer's liability is reduced or eliminated.

15. Type of damage

It is important to know what type of damage has occurred, whether injuries to persons or material damage or both. Not all types of damage fall within the Directive on product liability. For example, loss of profits or emotional damage can be compensated in the event of unlawful actions or under contract law (in the event of nonobservance of an agreement).

16. Limitation or expiration dates

Under both the Directive on product liability and other statutory (liability) regulations, dates are set for expiration and limitation of the claim. If the victim responds too late, the legal claim expires and no compensation can be paid.

17. Product recall

Besides the obligation to launch safe products on the market, the producer is also obliged to recall unsafe products. A producer who suspects that an unsafe product was launched on the market must decide whether to issue warnings or undertake a product recall.

18. Previous incidence of damage

If an incidence of damage has previously been caused by a comparable product, the producer can learn from this. There is probably a written record of the resolution of the incident.

19. Other suppliers

In certain cases the producer must investigate whether it is necessary to obtain damages for a wronged customer from other suppliers to satisfy a warranty claim.

Appendix

I

This appendix contains the text of the Machinery Directive 89/392/EEC and the text of the amendments, Directive 91/368/EEC and 93/44/EEC.

DIRECTIVE 89/392/EEC

Council Directive of 14 June 1989 on the approximation of the laws of the Member States relating to machinery (89/392/EEC).

THE COUNCIL OF THE EUROPEAN COMMUNITIES,

Having regard to the Treaty establishing the European Economic Community, and in particular Article 100a thereof,

Having regard to the proposal from the Commission,[1]

In cooperation with the European Parliament,[2]

Having regard to the opinion of the Economic and Social Committee,[3]

Whereas Member States are responsible for ensuring the health and safety on their territory of their people and, where appropriate, of domestic animals and goods and, in particular, of workers notably in relation to the risks arising out of the use of machinery;

Whereas, in the Member States, the legislative systems regarding accident prevention are very different; whereas the relevant compulsory provisions, frequently supplemented by defacto mandatory technical specifications and/or voluntary standards, do not necessarily lead to different levels of health and safety, but nevertheless, owing to their disparities, constitute barriers to trade within the Community; whereas, furthermore, conformity certification and national certification systems for machinery differ considerably;

---

[1] OJ No C 29, 3. 2. 1988, p. 1, and OJ No C 214, 16. 8. 1988, p. 23.
[2] OJ No C 326, 19. 12. 1988, p. 143, and OJ No C 158, 26. 6. 1989.
[3] OJ No C 337, 31. 12. 1988, p. 30.

Whereas the maintenance or improvement of the level of safety attained by the Member States constitutes one of the essential aims of this Directive and of the principle of safety as defined by the essential requirements;

Whereas existing national health and safety provisions providing protection against the risks caused by machinery must be approximated to ensure free movement of machinery without lowering existing justified levels of protection in the Member States; whereas the provisions of this Directive concerning the design and construction of machinery, essential for a safer working environment shall be accompanied by specific provisions concerning the prevention of certain risks to which workers can be exposed at work, as well as by provisions based on the organization of safety of workers in the working environment;

Whereas the machinery sector is an important part of the engineering industry and is one of the industrial mainstays of the Community economy;

Whereas paragraphs 65 and 68 of the White Paper on the completion of the internal market, approved by the European Council in June 1985, provide for a new approach to legislative harmonization;

Whereas the social cost of the large number of accidents caused directly by the use of machinery can be reduced by inherently safe design and construction of machinery and by proper installations and maintenance;

Whereas the field of application of this Directive must be based on a general definition of the term "machinery" so as to allow the technical development of products; whereas the development of "complex installations" and the risks they involve are of an equivalent nature and their express inclusion in the Directive is therefore justified;

Whereas specific Directives containing design and construction provisions for certain categories of machinery are now envisaged;

Whereas the very broad scope of this Directive must be limited in relation to these Directives and also existing Directives where they contain design and construction provisions;

Whereas Community law, in its present form, provides—by way of derogation from one of the fundamental rules of the Community, namely the free movement of goods—that obstacles to movement within the Community resulting from disparities in national legislation relating to the marketing of products must be accepted in so far as the provisions concerned can be recognized as being necessary to satisfy imperative requirements; whereas, therefore, the harmonization of laws in this case must be limited only to those requirements necessary to satisfy the imperative and essential health and safety requirements relating to machinery; whereas these requirements must replace the relevant national provisions because they are essential;

Whereas the essential health and safety requirements must be observed in order to ensure that machinery is safe; whereas these requirements must be applied with discernment to take account of the state of the art at the time of construction and of technical and economic requirements;

Whereas the putting into service of machinery within the meaning of this Directive can relate only to the use of the machinery itself as intended by the manufacturer; whereas this does not preclude the laying-down of conditions of use external to the machinery, provided that it is not thereby modified in a way not specified in this Directive;

Whereas, for trade fairs, exhibitions, etc., it must be possible to exhibit machinery which does not conform to this Directive; whereas, however, interested parties should be properly informed that the machinery does not conform and cannot be purchased in that condition;

Whereas, therefore, this Directive defines only the essential health and safety requirements of general application, supplemented by a number of more specific requirements for certain categories of machinery; whereas, in order to help manufacturers to prove conformity to these essential requirements and in order to allow inspection for conformity to the essential requirements, it is desirable to have standards harmonized at European level for the prevention of risks arising out of the design and construction of machinery; whereas these standards harmonized at European level are drawn up by private-law bodies and must retain their non-binding status; whereas for this purpose the European Committee for Standardization (CEN) and the European Committee for Electrotechnical Standardization (Cenelec) are the bodies recognized as competent to adopt harmonized standards in accordance with the general guidelines for cooperation between the Commission and these two bodies signed on 13 November 1984; whereas, within the meaning of this Directive, a harmonized standard is a technical specification (European standard or harmonization document) adopted by either or both of these bodies, on the basis of a remit from the Commission in accordance with the provisions of Council Directive 83/189/EEC of 28 March 1983 laying down a procedure for the provision of information in the field of technical standards and regulations,[4] as last amended by Directive 88/182/EEC,[5] and on the basis of general guidelines referred to above;

Whereas the legislative framework needs to be improved in order to ensure an effective and appropriate contribution by employers and employees to the standardization process; whereas such improvement should be completed at the latest by the time this Directive is implemented;

Whereas, as is currently the practice in Member States, manufacturers should retain the responsibility for certifying the conformity of their machinery to the relevant essential requirements; whereas conformity to harmonized standards creates a presumption of conformity to the relevant essential requirements; whereas it is left to the sole discretion of the manufacturer, where he feels the need, to have his products examined and certified by a third party;

Whereas, for certain types of machinery having a higher risk factor, a stricter certification procedure is desirable; whereas the EC type-examination procedure adopted may result in an EC declaration being given by the manufacturer without any stricter requirement such as a guarantee of quality, EC verification or EC supervision;

Whereas it is essential that, before issuing an EC declaration of conformity, the manufacturer or his authorized representative established in the Community should provide a technical construction file; whereas it is not, however, essential that all documentation be permanently available in a material manner but it must be made available on demand; whereas it need not include detailed plans of the sub-assemblies used in manufacturing the machines, unless knowledge of these is indispensable in order to ascertain conformity with essential safety requirements;

Whereas it is necessary not only to ensure the free movement and putting into service of machinery bearing the EC mark and having an EC conformity certificate but also to ensure free movement of machinery not bearing the EC mark where it is to be incorporated into other machinery or assembled with other machinery to form a complex installation;

---

[4] OJ No L 109, 26. 4. 1983, p. 8.
[5] OJ No L 81, 26. 3. 1988, p. 75

Whereas the Member States' responsibility for safety, health and the other aspects covered by the essential requirements on their territory must be recognized in a safeguard clause providing for adequate Community protection procedures;

Whereas the addresses of any decision taken under this Directive must be informed on the reasons for such a decision and the legal remedies open to them;

Whereas the measures aimed at the gradual establishment of the internal market must be adopted by 31 December 1992; whereas the internal market consists of an area without internal frontiers within which the free movement of goods, persons, services and capital is guaranteed,

HAS ADOPTED THIS DIRECTIVE:

CHAPTER I
SCOPE, PLACING ON THE MARKET AND FREEDOM OF MOVEMENT

ARTICLE 1

1. This Directive applies to machinery and lays down the essential health and safety requirements therefor, as defined in Annex I.

2. For the purposes of this Directive, "machinery" means an assembly of linked parts or components, at least one of which moves, with the appropriate actuators, control and power circuits, etc., joined together for a specific application, in particular for the processing, treatment, moving or packaging of a material.

The term "machinery" also covers an assembly of machines which, in order to achieve the same end, are arranged and controlled so that they function as an integral whole.

3. The following are excluded from the scope of this Directive:

- mobile equipment,
- lifting equipment,
- machinery whose only power source is directly applied manual effort,

- machinery for medical use used in direct contact with patients,
- special equipment for use in fairgrounds and/or amusement parks,
- steam boilers, tanks and pressure vessels,
- machinery specially designed or put into service for nuclear purposes which, in the event of failure, may result in an emission of radioactivity,
- radioactive sources forming part of a machine,
- firearms,
- storage tanks and pipelines for petrol, diesel fuel, inflammable liquids and dangerous substances.

4. Where, for machinery, the risks referred to in this Directive are wholly or partly covered by specific Community Directives, this Directive shall not apply, or shall cease to apply, in the case of such machinery and of such risks on the entry into force of these specific Directives.

Where, for machinery, the risks are mainly of electrical origin, such machinery shall be covered exclusively by Council Directive 73/23/EEC of 19 February 1973 on the harmonization of the laws of the Member States relating to electrical equipment designed for use within certain voltage limits.[6]

ARTICLE 2

1. Member States shall take all appropriate measures to ensure that machinery covered by this Directive may be placed on the market and put into service only if it does not endanger the health or safety of persons and, where appropriate, domestic animals or property, when properly installed and maintained and used for its intended purpose.

2. The provisions of this Directive shall not affect Member States' entitlement to lay down, in due observance of the Treaty, such requirements as they may deem necessary to ensure that persons and in particular workers are protected when using the machines in question, provided that this does not mean that the machinery is modified in a way not specified in the Directive.

---

[6] OJ No L 77, 26. 3. 1973, p. 29.

3. At trade fairs, exhibitions, demonstrations, etc., Member States shall not prevent the showing of machinery which does not conform to the provisions of this Directive, provided that a visible sign clearly indicates that such machinery does not conform and that it is not for sale until it has been brought into conformity by the manufacturer or his authorized representative established in the Community. During demonstrations, adequate safety measures shall be taken to ensure the protection of persons.

ARTICLE 3

Machinery covered by this Directive shall satisfy the essential health and safety requirements set out in Annex I.

ARTICLE 4

1. Member States shall not prohibit, restrict or impede the placing on the market and putting into service in their territory of machinery which complies with the provisions of this Directive.

2. Member States shall not prohibit, restrict or impede the placing on the market of machinery where the manufacturer or his authorized representative established in the Community declares in accordance with Annex II.B that it is intended to be incorporated into machinery or assembled with other machinery to constitute machinery covered by this Directive except where it can function independently.

ARTICLE 5

1. Member States shall regard machinery bearing the EC mark and accompanied by the EC declaration of conformity referred to in Annex II as conforming to the essential health and safety requirements referred to in Article 3.

In the absence of harmonized standards, Member States shall take any steps they deem necessary to bring to the attention of the parties concerned the existing national technical standards and specifications which are regarded as important or relevant to the proper implementation of the essential safety and health requirements in Annex I.

2. Where a national standard transposing a harmonized standard, the reference for which has been published in the Official Journal of the European Communities, covers one or more of the essential safety requirements, machinery constructed in accordance with this standard shall be presumed to comply with the relevant essential requirements.

Member States shall publish the references of national standards transposing harmonized standards.

3. Member States shall ensure that appropriate measures are taken to enable the social partners to have an influence at national level on the process of preparing and monitoring the harmonized standards.

ARTICLE 6

1. Where a Member State or the Commission considers that the harmonized standards referred to in Article 5, section 2, do not entirely satisfy the essential requirements referred to in Article 3, the Commission or the Member State concerned shall bring the matter before the Committee set up under Directive 83/189/EEC, giving the reasons therefor. The Committee shall deliver an opinion without delay.

Upon receipt of the Committee's opinion, the Commission shall inform the Member States whether or not it is necessary to withdraw those standards from the published information referred to in Article 5, section 2.

2. A standing committee shall be set up, consisting of representatives appointed by the Member States and chaired by a representative of the Commission.

The standing committee shall draw up its own rules of procedure.

Any matter relating to the implementation and practical application of this Directive may be brought before the standing committee, in accordance with the following procedure:

The representative of the Commission shall submit to the committee a draft of the measures to be taken. The committee shall deliver its opinion on the draft, within a time limit which the chairman may lay down according to the urgency of the matter, if necessary by taking a vote.

The opinion shall be recorded in the minutes; in addition, each Member State shall have the right to ask to have its position recorded in the minutes.

The Commission shall take the utmost account of the opinion delivered by the committee. It shall inform the committee of the manner in which its opinion has been taken into account.

ARTICLE 7

1. Where a Member State ascertains that machinery bearing the EC mark and used in accordance with its intended purpose is liable to endanger the safety of persons, and, where appropriate, domestic animals or property, it shall take all appropriate measures to withdraw such machinery from the market, to prohibit the placing on the market, putting into service or use thereof, or to restrict free movement thereof.

The Member State shall immediately inform the Commission of any such measure, indicating the reasons for its decision and, in particular, whether non-conformity is due to:

(a) failure to satisfy the essential requirements referred to in Article 3;
(b) incorrect application of the standards referred to in Article 5, section 2;
(c) shortcomings in the standards referred to in Article 5, section 2 themselves.

2. The Commission shall enter into consultation with the parties concerned without delay. Where the Commission considers, after this consultation, that the measure is justified, it shall immediately so inform the Member State which took the initiative and the other Member States. Where the Commission considers, after this consultation, that the action is unjustified, it shall immediately so inform the Member State which took the initiative and the manufacturer or his authorized representative established within the Community. Where the decision referred to in paragraph 1 is based on a shortcoming in the standards, and where the Member State at the origin of the decision maintains its position, the Commission shall immediately inform the Committee in order to initiate the procedures referred to in Article 6, section 1.

3. Where machinery which does not comply bears the EC mark, the competent Member State shall take appropriate action against whomso-

ever has affixed the mark and shall so inform the Commission and the other Member States.

4. The Commission shall ensure that the Member States are kept informed of the progress and outcome of this procedure.

CHAPTER II
CERTIFICATION PROCEDURE

ARTICLE 8

1. The manufacturer, or his authorized representative established in the Community, shall, in order to certify the conformity of machinery with the provisions of this Directive, draw up an EC declaration of conformity based on the model given in Annex II for each machine manufactured and shall affix to the machinery the EC mark referred to in Article 10.

2. Before placing on the market, the manufacturer, or his authorized representative established in the Community, shall:

(a) if the machinery is not referred to in Annex IV, draw up the file provided for in Annex V;

(b) if the machinery is referred to in Annex IV and its manufacturer does not comply, or only partly complies, with the standards referred to in Article 5, section 2, or if there are no such standards, submit an example of the machinery for the EC type-examination referred to in Annex VI;

(c) if the machinery is referred to in Annex IV and is manufactured in accordance with the standards referred to in Article 5, section 2:

- either draw up the file referred to in Annex VI and forward it to a notified body, which will acknowledge receipt of the file as soon as possible and keep it,
- submit the file referred to in Annex VI to the notified body, which will simply verify that the standards referred to in Article 5, section 2 have been correctly applied and will draw up a certificate of adequacy for the file,

- or submit the example of the machinery for the EC type-examination referred to in Annex VI.

3. Where the first indent of paragraph 2 (c) applies, the provisions of the first sentence of paragraph 5 and paragraph 7 of Annex VI shall also apply.

Where the second indent of 2 (c) applies, the provisions of paragraphs 5, 6 and 7 of Annex VI shall also apply.

4. Where paragraph 2 (a) and the first and second indents of paragraph 2 (c) apply, the EC declaration of conformity shall solely state conformity with the essential requirements of the Directive.

Where paragraph 2 (b) and (c) apply, the EC declaration of conformity shall state conformity with the example that underwent EC type-examination.

5. Where the machinery is subject to other Community Directives concerning other aspects, the EC mark referred to in Article 10 shall indicate in these cases that the machinery also fulfils the requirements of the other Directives.

6. Where neither the manufacturer nor his authorized representative established in the Community fulfils the obligations of the preceding paragraphs, these obligations shall fall to any person placing the machinery on the market in the Community. The same obligations shall apply to any person assembling machinery or parts thereof of various origins or constructing machinery for his own use.

ARTICLE 9

1. Each Member State shall notify the Commission and the other Member States of the approved bodies responsible for carrying out the certification procedures referred to in Article 8, section 2, (b) and (c). The Commission shall publish a list of these bodies in the Official Journal of the European Communities for information and shall ensure that the list is kept up to date.

2. Member States shall apply the criteria laid down in Annex VII in assessing the bodies to be indicated in such notification. Bodies meeting the assessment criteria laid down in the relevant harmonized standards shall be presumed to fulfil those criteria.

3. A Member State which has approved a body must withdraw its notification if it finds that the body no longer meets the criteria referred to in Annex VII. It shall immediately inform the Commission and the other Member States accordingly.

CHAPTER III
EC MARK

ARTICLE 10

1. The "EC" mark shall consist of the EC symbol followed by the last two digits of the year in which the mark was affixed.
Annex III shows the model to be used.
2. The EC mark shall be affixed to machinery distinctly and visibly in accordance with point 1.7.3 of Annex I.
3. Marks or inscriptions liable to be confused with the EC mark shall not be put on machinery.

CHAPTER IV
FINAL PROVISIONS

ARTICLE 11

Any decision taken pursuant to this Directive which restricts the marketing and putting into service of machinery shall state the exact grounds on which it is based. Such a decision shall be notified as soon as possible to the party concerned, who shall at the same time be informed of the legal remedies available to him under the laws in force in the Member State concerned and of the time limits to which such remedies are subject.

ARTICLE 12

The Commission will take the necessary steps to have information on all the relevant decisions relating to the management of this Directive made available.

ARTICLE 13

1. Member States shall adopt and publish the laws, regulations and administrative provisions necessary in order to comply with this Directive by 1 January 1992 at the latest. They shall forthwith inform the Commission thereof.

They shall apply these provisions with effect from 31 December 1992.

2. Member States shall ensure that the texts of the provisions of national law which they adopt in the field covered by this Directive are communicated to the Commission.

ARTICLE 14

This Directive is addressed to the Member States.
Done at Luxembourg, 14 June 1989.
For the Council
The President
P. SOLBES

ANNEX I
ESSENTIAL HEALTH AND SAFETY REQUIREMENTS RELATING TO THE DESIGN AND CONSTRUCTION OF MACHINERY

PRELIMINARY OBSERVATIONS

1. The obligations laid down by the essential health and safety requirements apply only when the corresponding hazard exists for the machinery in question when it is used under the conditions foreseen by the

manufacturer. In any event, requirements 1.1.2, 1.7.3 and 1.7.4 apply to all machinery covered by this Directive.

2. The essential health and safety requirements laid down in this Directive are mandatory. However, taking into account the state of the art, it may not be possible to meet the objectives set by them. In this case, the machinery must as far as possible be designed and constructed with the purpose of approaching those objectives.

1. ESSENTIAL HEALTH AND SAFETY REQUIREMENTS

1.1 General remarks

1.1.1 Definitions

For the purpose of this Directive

1. "danger zone" means any zone within and/or around machinery in which an exposed person is subject to a risk to his health or safety;

2. "exposed person" means any person wholly or partially in a danger zone;

3. "operator" means the person or persons given the task of installing, operating, adjusting, maintaining, cleaning, repairing or transporting machinery.

1.1.2 Principles of safety integration

(a) Machinery must be so constructed that it is fitted for its function, and can be adjusted and maintained without putting persons at risk when these operations are carried out under the conditions foreseen by the manufacturer.

The aim of measures taken must be to eliminate any risk of accident throughout the foreseeable lifetime of the machinery, including the phases of assembly and dismantling, even where risks of accident arise from foreseeable abnormal situations.

(b) In selecting the most appropriate methods, the manufacturer must apply the following principles, in the order given:
- eliminate or reduce risks as far as possible (inherently safe machinery design and construction),
- take the necessary protection measures in relation to risks that cannot be eliminated,

- inform users of the residual risks due to any shortcomings of the protection measures adopted, indicate whether any particular training is required and specify any need to provide personal protection equipment.

(c) When designing and constructing machinery, and when drafting the instructions, the manufacturer must envisage not only the normal use of the machinery but also uses which could reasonably be expected.

The machinery must be designed to prevent abnormal use if such use would engender a risk. In other cases the instructions must draw the user's attention to ways—which experience has shown might occur—in which the machinery should not be used.

(d) Under the intended conditions of use, the discomfort, fatigue and psychological stress faced by the operator must be reduced to the minimum possible taking ergonomic principles into account.

(e) When designing and constructing machinery, the manufacturer must take account of the constraints to which the operator is subject as a result of the necessary or foreseeable use of personal protection equipment (such as footwear, gloves, etc.).

(f) Machinery must be supplied with all the essential special equipment and accessories to enable it to be adjusted, maintained and used without risk.

1.1.3 Materials and products

The materials used to construct machinery or products used and created during its use must not endanger exposed persons' safety or health.

In particular, where fluids are used, machinery must be designed and constructed for use without risks due to filling, use, recovery or draining.

1.1.4 Lighting

The manufacturer must supply integral lighting suitable for the operations concerned where its lack is likely to cause a risk despite ambient lighting of normal intensity.

The manufacturer must ensure that there is no area of shadow likely to cause nuisance, that there is no irritating dazzle and that there are no dangerous stroboscopic effects due to the lighting provided by the manufacturer.

Internal parts requiring frequent inspection, and adjustment and maintenance areas, must be provided with appropriate lighting.

1.1.5 Design of machinery to facilitate its handling

Machinery or each component part thereof must:

- be capable of being handled safely,
- be packaged or designed so that it can be stored safely and without damage (e.g. adequate stability, special supports, etc.).

Where the weight, size or shape of machinery or its various component parts prevents them from being moved by hand, the machinery or each component part must:

- either be fitted with attachments for lifting gear, or
- be designed so that it can be fitted with such attachments (e.g. threaded holes), or
- be shaped in such a way that standard lifting gear can easily be attached.

Where machinery or one of its component parts is to be moved by hand, it must:

- either be easily movable, or
- be equipped for picking up (e.g. hand-grips, etc.) and moving in complete safety.

Special arrangements must be made for the handling of tools and/or machinery parts, even if lightweight, which could be dangerous (shape, material, etc.).

1.2 Controls

1.2.1 Safety and reliability of control systems

Control systems must be designed and constructed so that they are safe and reliable, in a way that will prevent a dangerous situation arising. Above all they must be designed and constructed in such a way that:

- they can withstand the rigours of normal use and external factors,
- errors in logic do not lead to dangerous situations.

1.2.2 Control devices

Control devices must be:

- clearly visible and identifiable and appropriately marked where necessary,
- positioned for safe operation without hesitation or loss of time, and without ambiguity,
- designed so that the movement of the control is consistent with its effect,
- located outside the danger zones, except for certain controls where necessary, such as emergency stop, console for training of robots,
- positioned so that their operation cannot cause additional risk,
- designed or protected so that the desired effect, where a risk is involved, cannot occur without an intentional operation,
- made so as to withstand foreseeable strain; particular attention must be paid to emergency stop devices liable to be subjected to considerable strain.

Where a control is designed and constructed to perform several different actions, namely where there is no one-to-one correspondence (e.g. keyboards, etc.), the action to be performed must be clearly displayed and subject to confirmation where necessary.

Controls must be so arranged that their layout, travel and resistance to operation are compatible with the action to be performed, taking account of ergonomic principles. Constraints due to the necessary or foreseeable use of personal protection equipment (such as footwear, gloves, etc.) must be taken into account.

Machinery must be fitted with indicators (dials, signals, etc.) as required for safe operation. The operator must be able to read them from the control position.

From the main control position the operator must be able to ensure that there are no exposed persons in the danger zones.

If this is impossible, the control system must be designed and constructed so that an acoustic and/or visual warning signal is given whenever the machinery is about to start. The exposed person must have the time and the means to take rapid action to prevent the machinery starting up.

1.2.3 Starting

It must be possible to start machinery only be voluntary actuation of a control provided for the purpose.

The same requirement applies:

- when restarting the machinery after a stoppage, whatever the cause,
- when effecting a significant change in the operating conditions (e.g. speed, pressure, etc.),

unless such restarting or change in operating conditions is without risk to exposed persons.

This essential requirement does not apply to the restarting of the machinery or to the change in operating conditions resulting from the normal sequence of an automatic cycle.

Where machinery has several starting controls and the operators can therefore put each other in danger, additional devices (e.g. enabling devices or selectors allowing only one part of the starting mechanism to be actuated at any one time) must be fitted to rule out such risks.

It must be possible for automated plant functioning in automatic mode to be restarted easily after a stoppage once the safety conditions have been fulfilled.

1.2.4 Stopping device

Normal stopping

Each machine must be fitted with a control whereby the machine can be brought safely to a complete stop.

Each workstation must be fitted with a control to stop some or all of the moving parts of the machinery, depending on the type of hazard, so that the machinery is rendered safe. The machinery's stop control must have priority over the start controls.

Once the machinery or its dangerous parts have stopped, the energy supply to the actuators concerned must be cut off.

Emergency stop

Each machine must be fitted with one or more emergency stop devices to enable actual or impending danger to be averted. The following exceptions apply:

- machines in which an emergency stop device would not lessen the risk, either because it would not reduce the stopping time or because it would not enable the special measures required to deal with the risk to be taken,
- hand-held portable machines and hand-guided machines.

This device must:

- have clearly identifiable, clearly visible and quickly accessible controls,
- stop the dangerous process as quickly as possible, without creating additional hazards,
- where necessary, trigger or permit the triggering of certain safeguard movements.

The emergency stop control must remain engaged; it must be possible to disengage it only by an appropriate operation; disengaging the control must not restart the machinery, but only permit restarting; the stop control must not trigger the stopping function before being in the engaged position.

Complex installations

In the case of machinery or parts of machinery designed to work together, the manufacturer must so design and construct the machinery that the stop controls, including the emergency stop, can stop not only the machinery itself but also all equipment upstream and/or downstream if its continued operation can be dangerous.

1.2.5 Mode selection

The control mode selected must override all other control systems with the exception of the emergency stop.

If machinery has been designed and built to allow for its use in several control or operating modes presenting different safety levels (e.g. to allow for adjustment, maintenance, inspection, etc.), it must be fitted with a mode selector which can be locked in each position. Each position of the selector must correspond to a single operating or control mode.

The selector may be replaced by another selection method which restricts the use of certain functions of the machinery to certain categories of operator (e.g. access codes for certain numerically controlled functions, etc.).

If, for certain operations, the machinery must be able to operate with its protection devices neutralized, the mode selector must simultaneously:

- disable the automatic control mode,
- permit movements only by controls requiring sustained action,
- permit the operation of dangerous moving parts only in enhanced safety conditions (e.g. reduced speed, reduced power, step-by-step, or other adequate provision) while preventing hazards from linked sequences,
- prevent any movement liable to pose a danger by acting voluntarily or involuntarily on the machine's internal sensors.

In addition, the operator must be able to control operation of the parts he is working on at the adjustment point.

1.2.6 Failure of the power supply

The interruption, re-establishment after an interruption or fluctuation in whatever manner of the power supply to the machinery must not lead to a dangerous situation.

In particular:

- the machinery must not start unexpectedly,
- the machinery must not be prevented from stopping if the command has already been given,
- no moving part of the machinery or piece held by the machinery must fall or be ejected,
- automatic or manual stopping of the moving parts whatever they may be must be unimpeded,
- the protection devices must remain fully effective.

1.2.7 Failure of the control circuit

A fault in the control circuit logic, or failure of or damage to the control circuit must not lead to dangerous situations.

In particular:

- the machinery must not start unexpectedly,
- the machinery must not be prevented from stopping if the command has already been given,
- no moving part of the machinery or piece held by the machinery must fall or be ejected,
- automatic or manual stopping of the moving parts whatever they may be must be unimpeded,
- the protection devices must remain fully effective.

1.2.8 Software

Interactive software between the operator and the command or control system of a machine must be user-friendly.

1.3 Protection against mechanical hazards

1.3.1 Stability

Machinery, components and fittings thereof must be so designed and constructed that they are stable enough, under the foreseen operating conditions (if necessary taking climatic conditions into account) for use without risk of overturning, falling or unexpected movement.

If the shape of the machinery itself or its intended installation does not offer sufficient stability, appropriate means of anchorage must be incorporated and indicated in the instructions.

1.3.2. Risk of break-up during operation

The various parts of machinery and their linkages must be able to withstand the stresses to which they are subject when used as foreseen by the manufacturer.

The durability of the materials used must be adequate for the nature of the work place foreseen by the manufacturer, in particular as regards the phenomena of fatigue, ageing, corrosion and abrasion.

The manufacturer must indicate in the instructions the type and frequency of inspection and maintenance required for safety reasons. The manufacturer must, where appropriate, indicate the parts subject to wear and the criteria for replacement.

Where a risk of rupture or disintegration remains despite the measures taken (e.g. as with grinding wheels) the moving parts must be mounted and positioned in such a way that in case of rupture their fragments will be contained.

Both rigid and flexible pipes carrying fluids, particularly those under high pressure, must be able to withstand the foreseen internal and external stresses and must be firmly attached and/or protected against all manner of external stresses and strains; precautions must be taken to ensure that no risk is posed by a rupture (sudden movement, high-pressure jets, etc.).

Where the material to be processed is fed to the tool automatically, the following conditions must be fulfilled to avoid risks to the persons exposed (e.g. tool breakage):

- when the workpiece comes into contact with the tool the latter must have attained its normal working conditions,
- when the tool starts and/or stops (intentionally or accidentally) the feed movement and the tool movement must be coordinated.

1.3.3 Risks due to falling or ejected objects

Precautions must be taken to prevent risks from falling or ejected objects (e.g. workpieces, tools, cuttings, fragments, waste, etc.).

1.3.4 Risks due to surfaces, edges or angles

In so far as their purpose allows, accessible parts of the machinery must have no sharp edges, no sharp angles, and no rough surfaces likely to cause injury.

1.3.5 Risks related to combined machinery

Where the machinery is intended to carry out several different operations with the manual removal of the piece between each operation (combined machinery), it must be designed and constructed in such a way as to enable each element to be used separately without the other elements constituting a danger or risk for the exposed person.

For this purpose, it must be possible to start and stop separately any elements that are not protected.

1.3.6 Risks relating to variations in the rotational speed of tools

When the machine is designed to perform operations under different conditions of use (e.g. different speeds or energy supply), it must be designed and constructed in such a way that selection and adjustment of these conditions can be carried out safely and reliably.

1.3.7 Prevention of risks related to moving parts

The moving parts of machinery must be designed, built and laid out to avoid hazards or, where hazards persist, fixed with guards or protective devices in such a way as to prevent all risk of contact which could lead to accidents.

1.3.8 Choice of protection against risks related to moving parts

Guards or protection devices used to protect against the risks related to moving parts must be selected on the basis of the type of risk. The following guidelines must be used to help make the choice.

A. Moving transmission parts
Guards designed to protect exposed persons against the risks associated with moving transmission parts (such as pulleys, belts, gears, rack and pinions, shafts, etc.) must be:

- either fixed, complying with requirements 1.4.1 and 1.4.2.1, or
- movable, complying with requirements 1.4.1 and 1.4.2.2.A.

Movable guards should be used where frequent access is foreseen.

B. Moving parts directly involved in the process
Guards or protection devices designed to protect exposed persons against the risks associated with moving parts contributing to the work

(such as cutting tools, moving parts of presses, cylinders, parts in the process of being machined, etc.) must be:

- wherever possible fixed guards complying with requirements 1.4.1 and 1.4.2.1,
- otherwise, movable guards complying with requirements 1.4.1 and 1.4.2.2.B or protection devices such as sensing devices (e.g. non-material barriers, sensor mats), remote-hold protection devices (e.g. two-hand controls), or protection devices intended automatically to prevent all or part of the operator's body from encroaching on the danger zone in accordance with requirements 1.4.1 and 1.4.3.

However, when certain moving parts directly involved in the process cannot be made completely or partially inaccessible during operation owing to operations requiring nearby operator intervention, where technically possible such parts must be fitted with:

- fixed guards, complying with requirements 1.4.1 and 1.4.2.1 preventing access to those sections of the parts that are not used in the work,
- adjustable guards, complying with requirements 1.4.1 and 1.4.2.3

restricting access to those sections of the moving parts that are strictly for the work.

1.4 Required characteristics of guards and protection devices

1.4.1 General requirement

Guards and protection devices must:

- be of robust construction,
- not give rise to any additional risk,
- not be easy to by-pass or render non-operational,
- be located at an adequate distance from the danger zone,
- cause minimum obstruction to the view of the production process,
- enable essential work to be carried out on installation and/or replacement of tools and also for maintenance by restricting

access only to the area where the work has to be done, if possible without the guard or protection device having to be dismantled.

1.4.2 Special requirements for guards

1.4.2.1 Fixed guards
Fixed guards must be securely held in place.
They must be fixed by systems that can be opened only with tools.
Where possible, guards must be unable to remain in place without their fixings.

1.4.2.2 Movable guards
A. Type A movable guards must:

- as far as possible remain fixed to the machinery when open,
- be associated with a locking device to prevent moving parts starting up as long as these parts can be accessed and to give a stop command whenever they are no longer closed.

B. Type B movable guards must be designed and incorporated into the control system so that:

- moving parts cannot start up while they are within the operator's reach,
- the exposed person cannot reach moving parts once they have started up,
- they can be adjusted only by means of an intentional action, such as the use of a tool, key, etc.,
- the absence or failure of one of their components prevents starting or stops the moving parts,
- protection against any risk of ejection is proved by means of an appropriate barrier.

1.4.2.3 Adjustable guards restricting access
Adjustable guards restricting access to those areas of the moving parts strictly necessary for the work must:

- be adjustable manually or automatically according to the type of work involved,
- be readily adjustable without the use of tools,
- reduce as far as possible the risk of ejection.

1.4.3 Special requirements for protection devices

Protection devices must be designed and incorporated into the control system so that:
- moving parts cannot start up while they are within the operator's reach,
- the exposed person cannot reach moving parts once they have started up,
- they can be adjusted only be means of an intentional action, such as the use of a tool, key, etc.,
- the absence or failure of one of their components prevents starting or stops the moving parts.

1.5 Protection against other hazards

1.5.1 Electricity supply

Where machinery has an electricity supply it must be designed, constructed and equipped so that all hazards of an electrical nature are or can be prevented.

The specific rules in force relating to electrical equipment designed for use within certain voltage limits must apply to machinery which is subject to those limits.

1.5.2 Static electricity

Machinery must be so designed and constructed as to prevent or limit the build-up of potentially dangerous electrostatic charges and/or be fitted with a discharging system.

1.5.3 Energy supply other than electricity

Where machinery is powered by an energy other than electricity (e.g. hydraulic, pneumatic or thermal energy, etc.), it must be so designed, constructed and equipped as to avoid all potential hazards associated with these types of energy.

1.5.4 Errors of fitting

Errors likely to be made when fitting or refitting certain parts which could be a source of risk must be made impossible by the design of such parts or, failing this, by information given on the parts themselves and/or the housings. The same information must be given on moving parts and/or their housings where the direction of movement must be known to avoid a risk. Any further information that may be necessary must be given in the instructions.

Where a faulty connection can be the source of risk, incorrect fluid connections, including electrical conductors, must be made impossible by the design or, failing this, by information given on the pipes, cables, etc. and/or connector blocks.

1.5.5 Extreme temperatures

Steps must be taken to eliminate any risk of injury caused by contact with or proximity to machinery parts or materials at high or very low temperatures.

The risk of hot or very cold material being ejected should be assessed. Where this risk exists, the necessary steps must be taken to prevent it or, if this is not technically possible, to render it non-dangerous.

1.5.6 Fire

Machinery must be designed and constructed to avoid all risk of fire or overheating posed by the machinery itself or by gases, liquids, dusts, vapors or other substances produced or used by the machinery.

1.5.7 Explosion

Machinery must be designed and constructed to avoid any risk of explosion posed by the machinery itself or by gases, liquids, dusts, vapors or other substances produced or used by the machinery.

To that end the manufacturer must take steps to:

- avoid a dangerous concentration of products,

- prevent combustion of the potentially explosive atmosphere,
- minimize any explosion which may occur so that it does not endanger the surroundings.

The same precautions must be taken if the manufacturer foresees the use of the machinery in a potentially explosive atmosphere.

Electrical equipment forming part of the machinery must conform, as far as the risk from explosion is concerned, to the provision of the specific Directives in force.

1.5.8 Noise

Machinery must be so designed and constructed that risks resulting from the emission of airborne noise are reduced to the lowest level taking account of technical progress and the availability of means of reducing noise, in particular at source.

1.5.9 Vibration

Machinery must be so designed and constructed that risks resulting from vibrations produced by the machinery are reduced to the lowest level, taking account of technical progress and the availability of means of reducing vibration, in particular at source.

1.5.10 Radiation

Machinery must be so designed and constructed that any emission of radiation is limited to the extent necessary for its operation and that the effects on exposed persons are non-existent or reduced to non-dangerous proportions.

1.5.11 External radiation

Machinery must be so designed and constructed that external radiation does not interfere with its operation.

1.5.12 Laser equipment

Where laser equipment is used, the following provisions should be taken into account:

- laser equipment on machinery must be designed and constructed so as to prevent any accidental radiation,
- laser equipment on machinery must be protected so that effective radiation, radiation produced by reflection or diffusion and secondary radiation do not damage health,
- optical equipment for the observation or adjustment of laser equipment on machinery must be such that no health risk is created by the laser rays.

1.5.13 Emissions of dust, gases, etc.

Machinery must be so designed, constructed and/or equipped that risks due to gases, liquids, dust, vapors and other waste materials which it produces can be avoided.

Where a hazard exists, the machinery must be so equipped that the said substances can be contained and/or evacuated.

Where machinery is not enclosed during normal operation, the devices for containment and/or evacuation must be situated as close as possible to the source emission.

1.6 Maintenance

1.6.1 Machinery maintenance

Adjustment, lubrication and maintenance points must be located outside danger zones. It must be possible to carry out adjustment, maintenance, repair, cleaning and servicing operations while machinery is at a standstill.

If one or more of the above conditions cannot be satisfied for technical reasons, these operations must be possible without risk (see 1.2.5).

In the case of automated machinery and, where necessary, other machinery, the manufacturer must make provision for a connecting device for mounting diagnostic fault-finding equipment.

Automated machine components which have to be changed frequently, in particular for a change in manufacture or where they are liable to wear or likely to deteriorate following an accident, must be capable of being removed and replaced easily and in safety. Access to the components must enable these tasks to be carried out with the necessary technical means (tools, measuring instruments, etc.) in accordance with an operating method specified by the manufacturer.

1.6.2 Access to operating position and servicing points

The manufacturer must provide means of access (stairs, ladders, catwalks, etc.) to allow access in safety to all areas used for production, adjustment and maintenance operations.

Parts of the machinery where persons are liable to move about or stand must be designed and constructed to avoid falls.

1.6.3 Isolation of energy sources

All machinery must be fitted with means to isolate it from all energy sources. Such isolators must be clearly identified. They must be capable of being locked if reconnection could endanger exposed persons. In the case of machinery supplied with electricity through a plug capable of being plugged into a circuit, separation of the plug is sufficient.

The isolator must be capable of being locked also where an operator is unable, from any of the points to which he has access, to check that the energy is still cut off.

After the energy is cut off, it must be possible to dissipate normally any energy remaining or stored in the circuits of the machinery without risk to exposed persons.

As an exception to the above requirements, certain circuits may remain connected to their energy sources in order, for example, to hold parts, protect information, light interiors, etc. In this case, special steps must be taken to ensure operator safety.

1.6.4 Operator intervention

Machinery must be so designed, constructed and equipped that the need for operator intervention is limited.

If operator intervention cannot be avoided, it must be possible to carry it out easily and in safety.

1.7 Indicators

1.7.0 Information devices

The information needed to control machinery must be unambiguous and easily understood. It must not be excessive to the extent of overloading the operator.

1.7.1 Warning devices

Where machinery is equipped with warning devices (such as signals, etc.), these must be unambiguous and easily perceived.

The operator must have facilities to check the operation of such warning devices at all times.

The requirements of the specific Directives concerning colours and safety signals must be complied with.

1.7.2 Warning of residual risks

Where risks remain despite all the measures adopted or in the case of potential risks which are not evident (e.g. electrical cabinets, radioactive sources, bleeding of a hydraulic circuit, hazard in an unseen area, etc.), the manufacturer must provide warnings.

Such warnings should preferably use readily understandable pictograms and/or be drawn up in one of the languages of the country in which the machinery is to be used, accompanied, on request, by the languages understood by the operators.

1.7.3 Marking

All machinery must be marked legibly and indelibly with the following minimum particulars:

- name and address of the manufacturer,
- EC mark, which includes the year of construction (see Annex III),
- designation of series or type,
- serial number, if any.

Furthermore, where the manufacturer constructs machinery intended for use in a potentially explosive atmosphere, this must be indicated on the machinery.

Machinery must also bear full information relevant to its type and essential to its safe use (e.g. maximum speed of certain rotating parts, maximum diameter of tools to be fitted, mass, etc.).

1.7.4 Instructions

(a) All machinery must be accompanied by instructions including at least the following:

- a repeat of the information with which the machinery is marked (see 1.7.3), together with any appropriate additional information to facilitate maintenance (e.g. addresses of the importer, repairers, etc.),
- foreseen use of the machinery within the meaning of 1.1.2 (c),
- workstation(s) likely to be occupied by operators,
- instructions for safe:
  - putting into service,
  - use,
  - handling, giving the mass of the machinery and its various parts where they are regularly to be transported separately,
  - assembly, dismantling,
  - adjustment,
  - maintenance (servicing and repair),
- where necessary, training instructions.

Where necessary, the instructions should draw attention to ways in which the machinery should not be used.

(b) The instructions must be drawn up by the manufacturer or his authorized representative established in the Community in one of the languages of the country in which the machinery is to be used and should preferably be accompanied by the same instructions drawn up in another Community language, such as that of the country in which the manufacturer or his authorized representative is established. By way of derogation from this requirement, the maintenance instructions for use by the specialized personnel frequently employed by the manufacturer or his authorized representative may be drawn up in only one of the official Community languages.

(c) The instructions must contain the drawings and diagrams necessary for putting into service, maintenance, inspection, checking of correct operation and, where appropriate, repair of the machinery, and all useful instructions in particular with regard to safety.

(d) Any sales literature describing the machinery must not contradict the instructions as regards safety aspects; it must give information regarding the airborne noise emissions referred to in (f) and, in the case of hand-held and/or hand-guided machinery, information regarding vibration as referred to in 2.2.

(e) Where necessary, the instructions must give the requirements relating to installation and assembly for reducing noise or vibration (e.g. use of dampers, type and mass of foundation block, etc.).

(f) The instructions must give the following information concerning airborne noise emissions by the machinery, either the actual value or a value established on the basis of measurements made on identical machinery:

- equivalent continuous A-weighted sound pressure level at workstations, where this exceeds 70 dB(A); where this level does not exceed 70 dB(A), this fact must be indicated,
- peak C-weighted instantaneous sound pressure value at workstations, where this exceeds 63 Pa (130 dB in relation to 20 µPa),
- sound power level emitted by the machinery where the equivalent continuous A-weighted sound pressure level at workstations exceeds 85 dB(A).

In the case of very large machinery, instead of the sound power level, the equivalent continuous sound pressure levels at specified positions around the machinery may be indicated.

Sound levels must be measured using the most appropriate method for the machinery.

The manufacturer must indicate the operating conditions of the machinery during measurement and what methods have been used for the measurement.

Where the workstation(s) are undefined or cannot be defined, sound pressure levels must be measured at a distance of 1 metre from the surface of the machinery and at height of 1.60 metres from the floor or access platform. The position and value of the maximum sound pressure must be indicated.

(g) If the manufacturer foresees that the machinery will be used in a potentially explosive atmosphere, the instructions must give all the necessary information.

(h) In the case of machinery which may also be intended for use by non-professional operators, the wording and layout of the instructions for use, whilst respecting the other essential requirements mentioned above, must take into account the level of general education and acumen that can reasonably be expected from such operators.

2. ADDITIONAL ESSENTIAL HEALTH AND SAFETY REQUIREMENTS FOR CERTAIN CATEGORIES OF MACHINERY

2.1 Agri-foodstuffs machinery

In addition to the essential health and safety requirements set out in 1 above, where machinery is intended to prepare and process foodstuffs (e.g. cooking, refrigeration, thawing, washing, handling, packaging, storage, transport or distribution), it must be so designed and constructed as to avoid any risk of infection, sickness or contagion and the following hygiene rules must be observed:

(a) materials in contact, or intended to come into contact, with the foodstuffs must satisfy the conditions set down in the relevant Directives. The machinery must be so designed and constructed that these materials can be clean before each use;

(b) all surfaces including their joinings must be smooth, and must have neither ridges nor crevices which could harbour organic materials;

(c) assemblies must be designed in such a way as to reduce projections, edges and recesses to a minimum. They should preferably be made by welding or continuous bonding. Screws, screwheads and rivets may not be used except where technically unavoidable;

(d) all surfaces in contact with the foodstuffs must be easily cleaned and disinfected, where possible after removing easily dismantled parts. The inside surfaces must have curves of a radius sufficient to allow thorough cleaning;

(e) liquid deriving from foodstuffs as well as cleaning, disinfecting and rinsing fluids should be able to be discharged from the machine without impediment (possibly in a "clean" position);

(f) machinery must be so designed and constructed as to prevent any liquids or living creatures, in particular insects, entering, or any organic matter accumulating in areas that cannot be cleaned (e.g. for machinery not mounted on feet or casters, by placing a seal between the machinery and its base, by the use of sealed units, etc.);

(g) machinery must be so designed and constructed that no ancillary substances (e.g. lubricants, etc.) can come into contact with foodstuffs. Where necessary, machinery must be designed and constructed so that continuing compliance with this requirement can be checked.

Instructions

In addition to the information required in section 1, the instructions must indicate recommended products and methods for cleaning, disinfecting and rinsing (not only for easily accessible areas but also where areas to which access is impossible or unadvisable, such as piping, have to be cleaned in situ).

2.2 Portable hand-held and/or hand-guided machinery

In addition to the essential health and safety requirements set out in 1 above, portable hand-held and/or hand-guided machinery must conform to the following essential health and safety requirements:

- according to the type of machinery, it must have a supporting surface of sufficient size and have a sufficient number of handles and supports of an appropriate size and arranged to ensure the stability of the machinery under the operating conditions foreseen by the manufacturer,
- except where technically impossible or where there is an independent control, in the case of handles which cannot be released on complete safety, it must be fitted with start and stop controls arranged in such a way that the operator can operate them without releasing the handles,
- it must be designed, constructed or equipped to eliminate the risks of accidental starting and/or continued operation after the operator has released the handles. Equivalent steps must be taken if this requirement is not technically feasible,
- portable hand-held machinery must be designed and constructed to allow, where necessary, a visual check of the contact of the tool with the material being processed.

Instructions

The instructions must give the following information concerning vibrations transmitted by hand-held and hand-guided machinery:

- the weighted root mean square acceleration value to which the arms are subjected, if it exceeds 2.5 m/s^2 as determined by the appropriate test code. Where the acceleration does not exceed 2.5 m/s², this must be mentioned.

If there is no applicable test code, the manufacturer must indicate the measurement methods and conditions under which measurements were made.

2.3 Machinery for working wood and analogous materials

In addition to the essential health and safety requirements set out in 1 above, machinery for working wood and machinery for working materials with physical and technological characteristics similar to those of wood, such as cork, bone, hardened rubber, hardened plastic material and other similar stiff material must conform to the following essential health and safety requirements:

(a) the machinery must be designed, constructed or equipped so that the piece being machined can be placed and guided in safety; where the piece is hand-held on a work-bench the latter must be sufficiently stable during the work and must not impede the movement of the piece;

(b) where the machinery is likely to be used in conditions involving the risk of ejection of pieces of wood, it must be designed, constructed or equipped to eliminate this ejection, or, if this is not the case, so that the ejection does not engender risks for the operator and/or exposed persons;

(c) the machinery must be equipped with an automatic brake that stops the tool in a sufficiently short time if there is a risk of contact with the tool whilst it runs down;

(d) where the tool is incorporated into a non-fully automated machine, the latter must be so designed and constructed as to eliminate or reduce the risk of serious accidental injury, for example by using cylindrical cutter blocks, restricting depth of cut, etc.

ANNEX II

A. Contents of the EC declaration of conformity.[7]

---

[7] This declaration must be drawn up in the same language as the instructions (see Annex I, point 1.7.4) and must be either typewritten or handwritten in block capitals.

The EC declaration of conformity must contain the following particulars:

- name and address of the manufacturer or his authorized representative established in the Community.[8]
- description of the machinery.[9]
- all relevant provisions complied with by the machinery,
- where appropriate, name and address of the notified body and number of the EC type-examination certificate,
- where appropriate, the name and address of the notified body to which the file has been forwarded in accordance with the first indent of Article 8, section 2 (c),
- where appropriate, the name and address of the notified body which has carried out the verification referred to in the second indent of Article 8, section 2 (c),
- where appropriate, a reference to the harmonized standards,
- where appropriate, the national technical standards and specifications used,
- identification of the person empowered to sign on behalf of the manufacturer or his authorized representatives.

B. Contents of the declaration by the manufacturer or his authorized representatives established in the Community (Article 4, section 2)

The manufacturer's declaration referred to in Article 4, section 2 must contain the following particulars:

- name and address of the manufacturer or the authorized representative,
- description of the machinery or machinery parts,
- a statement that the machinery must not be put into service until the machinery into which it is to be incorporated has been declared in conformity with the provisions of the Directive,
- identification of the person signing.

---

[8] Business name and full address; authorized representatives must also give the business name and address of the manufacturer.
[9] Description of the machinery (make, type, serial number, etc.).

ANNEX III
EC MARK

The EC mark consists of the symbol shown below and the last two figures of the year in which the mark was affixed.

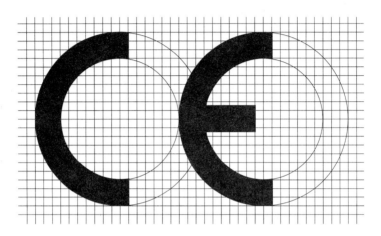

The different elements of the EC mark should have materially the same vertical dimensions, which should not be less than 5 mm.

ANNEX IV
TYPES OF MACHINES FOR WHICH THE PROCEDURE REFERRED TO IN ARTICLE 8, SECTION 2 (B) AND (C) MUST BE APPLIED

1. Circular saws (single-or multi-blade) for working with wood and meat.

1.1 Sawing machines with fixed tool during operation, having a fixed bed with manual feed of the workpiece or with a demountable power feed.

1.2 Sawing machines with fixed tool during operation, having a manually operated reciprocating saw-bench or carriage.

1.3 Sawing machines with fixed tool during operation, having a built-in mechanical feed device for the workpieces, with manual loading and/or unloading.

1.4 Sawing machines with movable tool during operation, with a mechanical feed device and manual loading and/or unloading.

2. Hand-fed surface planing machines for woodworking.

3. Thicknessers for one-side dressing with manual loading and/or unloading for woodworking.

4. Band-saws with a mobile bed or carriage and manual loading and/or unloading for working with wood and meat.

5. Combined machines of the types referred to in 1 to 4 and 7 for woodworking

6. Hand-fed tenoning machines with several tool holders for woodworking.

7. Hand-fed vertical spindle moulding machines.

8. Portable chain saws for woodworking.

9. Presses, including press-brakes, for the cold working of metals, with manual loading and/or unloading, whose movable working parts may have a travel exceeding 6 mm and a speed exceeding 30 mm/s.

10. Injection or compression plastics-moulding machines with manual loading or unloading.

11. Injection or compression rubber-moulding machines with manual loading or unloading.

12. Cartridge-operated fixing guns.

ANNEX V
EC DECLARATION OF CONFORMITY

1. The EC declaration of conformity is the procedure by which the manufacturer, or his authorized representative established in the Community declares that the machinery being placed on the market complies with all the essential health and safety requirements applying to it.

2. Signature of the EC declaration of conformity authorizes the manufacturer, or his authorized representative in the Community, to affix the EC mark to the machinery.

3. Before drawing up the EC declaration of conformity, the manufacturer, or his authorized representative in the Community, shall have ensured and be able to guarantee that the documentation listed below is and will remain available on his premises for any inspection purposes:

(a) a technical construction file comprising:

- an overall drawing of the machinery together with drawings of the control circuits,
- full detailed drawings, accompanied by any calculation notes, test results, etc., required to check the conformity of the machinery with the essential health and safety requirements,
- a list of:
    - the essential requirements of this Directive,
    - standards, and
    - other technical specifications, which were used when the machinery was designed,
- a description of methods adopted to eliminate hazards presented by the machinery,
- if he so desires, any technical report or certificate obtained from a competent body or laboratory.[10]
- if he declares conformity with a harmonized standard which provides therefore, any technical report giving the results of tests carried out at his choice either by himself or by a competent body or laboratory,[11]
- a copy of the instructions for the machinery;

(b) for series manufacture, the internal measures that will be implemented to ensure that the machinery remains in conformity with the provisions of the Directive.

The manufacturer must carry out necessary research or tests on components, fittings or the completed machine to determine whether by its design or construction, the machine is capable of being erected and put into service safely.

Failure to present the documentation in response to a duly substantiated request by the competent national authorities may constitute sufficient grounds for doubting the presumption of conformity with the requirements of the Directive.

---

[10] A body or laboratory is presumed competent if it meets the assessment criteria laid down in the relevant harmonized standards.

[11] See note 10.

4. (a) The documentation referred to in 3 above need not permanently exist in a material manner but it must be possible to assemble it and make it available within a period of time commensurate with its importance. It does not have to include detailed plans or any other specific information as regards the sub-assemblies used for the manufacture of the machinery unless a knowledge of them is essential for verification of conformity with the basic safety requirements.

(b) The documentation referred to in 3 above shall be retained and kept available for the competent national authorities for at least 10 years following the date of manufacture of the machinery or of the last unit produced, in the case of series manufacture.

(c) The documentation referred to in 3 above shall be drawn up in one of the official languages of the Communities, with the exception of the instructions for the machinery.

ANNEX VI
EC TYPE-EXAMINATION

1. EC type-examination is the procedure by which a notified body ascertains and certifies that an example of machinery satisfies the provisions of this Directive which apply to it.

2. The application for EC type-examination shall be lodged by the manufacturer or by his authorized representative established in the Community, with a single notified body in respect of an example of the machinery.

The application shall include:

- the name and address of the manufacturer or his authorized representative established in the Community and the place of manufacture of the machinery,
- a technical file comprising at least:
    - an overall drawing of the machinery together with drawings of the control circuits,
    - full detailed drawings, accompanied by any calculation notes, test results, etc., required to check the conformity of the machinery with the essential health and safety requirements,

- a description of methods adopted to eliminate hazards presented by the machinery and a list of standards used,
- a copy of the instructions for the machinery,
- for series manufacture, the internal measures that will be implemented to ensure that the machinery remains in conformity with the provisions of the Directive.

It shall be accompanied by a machine representative of the production planned or, where appropriate, a statement of where the machine may be examined

The documentation referred to above does not have to include detailed plans or any other specific information as regards the sub-assemblies used for the manufacture of the machinery unless a knowledge of them is essential for verification of conformity with the basic safety requirements.

3. The notified body shall carry out the EC type-examination in the manner described below:

- it shall examine the technical construction file to verify its appropriateness and the machine supplied or made available to it.
- during the examination of the machine, the body shall

(a) ensure that it has been manufactured in conformity which the technical construction file and may safely be used under its intended working conditions;

(b) check that standards, if used, have been properly applied;

(c) perform appropriate examinations and tests to check that the machine complies with the essential health and safety requirements applicable to it.

4. If the example complies with the provisions applicable to it the body shall draw up an EC type-examination certificate which shall be forwarded to the applicant. That certificate shall state the conclusions of the examination, indicate any conditions to which its issue may be subject and be accompanied by the descriptions and drawings necessary for identification of the approved example.

The Commission, the Member States and the other approved bodies may obtain a copy of the certificate and, on a reasoned request, a copy of the technical construction file and of the reports on the examinations and tests carried out.

5. The manufacturer or his authorized representative established in the Community shall inform the notified body of any modifications, even of a minor nature, which he has made or plans to make to the machine to which the example relates. The notified body shall examine those modifications and inform the manufacturer or his authorized representative established in the Community whether the EC type-examination certificate remains valid.

6. A body which refuses to issue an EC type-examination certificate shall so inform the other notified bodies. A body which withdraws an EC type-examination certificate shall so inform the Member State which notified it. The latter shall inform the other Member States and the Commission thereof, giving the reasons for the decision.

7. The files and correspondence referring to the EC type-examination procedures shall be drawn up in an official language of the Member State where the notified body is established or in a language acceptable to it.

ANNEX VII
MINIMUM CRITERIA TO BE TAKEN INTO ACCOUNT BY MEMBER STATES FOR THE NOTIFICATION OF BODIES

1. The body, its director and the staff responsible for carrying out the verification tests shall not be the designer, manufacturer, supplier or installer of machinery which they inspect, nor the authorized representative of any of these parties. They shall not become either involved directly or as authorized representatives in the design, construction, marketing or maintenance of the machinery. This does not preclude the possibility of exchanges of technical information between the manufacturer and the body.

2. The body and its staff shall carry out the verification tests with the highest degree of professional integrity and technical competence and shall be free from all pressures and inducements, particularly financial, which might influence their judgement or the results of the inspection, especially from persons or groups of persons with an interest in the result of verifications.

3. The body shall have at its disposal the necessary staff and possess the necessary facilities to enable it to perform properly the administra-

tive and technical tasks connected with verification; it shall also have access to the equipment required for special verification.

4. The staff responsible for inspection shall have:

- sound technical and professional training,
- satisfactory knowledge of the requirements of the tests they carry out and adequate experience of such tests,
- the ability to draw up the certificates, records and reports required to authenticate the performance of the tests.

5. The impartiality of inspection staff shall be guaranteed. Their remuneration shall not depend on the number of tests carried out or on the results of such tests.

6. The body shall take out liability insurance unless its liability is assumed by the State in accordance with national law, or the Member State itself is directly responsible for the tests.

7. The staff of the body shall be bound to observe professional secrecy with regard to all information gained in carrying out its tasks (except *vis-à-vis* the competent administrative authorities of the State in which its activities are carried out) under this Directive or any provision of national law giving effect to it.

CORRIGENDUM

Corrigendum to Council Directive 89/392/EEC of 14 June 1989 on the approximation of the laws of the Member States relating to machinery

(Official Journal of the European Communities No L 183 of 29 June 1989)

On page 13, Article 8, section 4, the second subparagraph shall read as follows:

"Where paragraph 2 (b) and the third indent of paragraph 2 (c) apply, the EC declaration of conformity shall state conformity with the example that underwent EC type-examination."

DIRECTIVE 91/368/EEC

Council Directive of 20 June 1991 amending Directive 89/392/EEC on the approximation of the laws of the Member States relating to machinery (91/368/EEC)

THE COUNCIL OF THE EUROPEAN COMMUNITIES,
Having regard to the Treaty establishing the European Economic Community, and in particular Article 100a thereof,
Having regard to the proposal from the Commission,[1]
In cooperation with the European Parliament.[2]
Having regard to the opinion of the Economic and Social Committee.[3]

Whereas machinery entailing specific risks due either to its mobility or its ability to lift loads, or to both these phenomena together, must satisfy both the general health and safety requirements set out in Directive 89/392/EEC[4] and the health and safety requirements relating to those specific risks;

Whereas it is unnecessary to provide for certification procedures for those types of machinery other than those initially provided for in Directive 89/392/EEC.

Whereas prescribing supplementary essential health and safety requirements for the specific risks due to mobility and the lifting of loads can be effected by amending Directive 89/392/EEC to include these complementary provisions; whereas this amendment can be used to correct certain imperfections in the essential health and safety requirements applying to all machinery;

Whereas it is necessary to provide for transitional arrangements enabling Member States to authorize the placing on the market and putting into service of machinery manufactured in accordance with the national rules in force on 31 December 1992;

---

[1] OJ No C 37, 17. 2. 1990, p. 5; and OJ No C 268, 24. 10. 1990, p. 12.
[2] OJ No C 175, 16.7. 1990, p. 119; and OJ No C 129, 20. 5. 1991.
[3] OJ No C 168, 10. 7. 1990, p. 15.
[4] OJ No L 183, 29. 6. 1989, p. 9

Whereas certain equipment or machinery covered by existing Directives comes within the scope of this Directive; whereas it is preferable to have one single Directive to cover all equipment; whereas it is therefore desirable that the relevant existing Directives be repealed on the date this Directive is applied,

HAS ADOPTED THIS DIRECTIVE:

ARTICLE 1

Directive 89/392/EEC is hereby amended as follows:

1. Article 1 is amended as follows:

(a) in paragraph 2, the following subparagraph shall be added:

"'Machinery' also means interchangeable equipment modifying the function of a machine, which is placed on the market for the purpose of being assembled with a machine or a series of different machines or with a tractor by the operator himself in so far as this equipment is not a spare part or a tool";

(b) in paragraph 3:

- the first indent is deleted,
- the second indent is replaced by the following: "- lifting equipment designed and constructed for raising and/or moving persons with or without loads, except for industrial trucks with elevating operator position."
- in the third indent, the following phrase is added:

    "unless it is a machine used for lifting or lowering loads,"

- the following indents are added:

    "• means of transport, i.e. vehicles and their trailers intended solely for transporting passengers by air or on road, rail or water networks, as well as means of transport in so far as such means are designed for transporting goods by air, on public road or rail networks or on water. Vehicles used in the mineral extraction industry shall not be excluded,
    - seagoing vessels and mobile offshore units together with equipment on board such vessels or units,
    - cableways for the public or private transportation of persons,
    - agricultural and forestry tractors, as defined in Article 1, section 1, of Council Directive 74/150/EEC of 4 March 1974 on the approximation of the laws of the Member States relating to the type-approval of

wheeled agricultural or forestry tractors[5], as lasted amended by Directive 88/297/EEC.[6]
- machines specially designed and constructed for military or police purposes."

2. In Article 2, section 3, first sentence, the phrase "... provisions of this Directive" is replaced by "Community provisions in force."

3. In Article 4, section 2, the following subparagraph is added:

"'Interchangeable equipment,' within the meaning of the third subparagraph of Article 1, section 2, shall be regarded as machinery and accordingly must in all cases bear the EC mark and be accompanied by the EC declaration of conformity referred to in Annex II (A)."

4. In Article 8, the following paragraph is added:

"7. The obligations laid down in paragraph 6 shall not apply to persons who assemble with a machine or tractor interchangeable equipment as provided for in Article 1, provided that the parts are compatible and each of the constituent parts of the assembled machine bears the EC mark and is accompanied by the EC declaration of conformity."

5. Article 13 is replaced by the following:

"Article 13

1. Before 1 January 1992 Member States shall adopt and publish the laws, regulations and administrative provisions necessary in order to comply with this Directive. They shall forthwith inform the Commission thereof.

When Member States adopt these measures, they shall contain a reference to this Directive or shall be accompanied by such reference on the occasion of their official publication. The methods of making such a reference shall be laid down by the Member States.

The Member States shall apply the measures in question with effect from 1 January 1993, except as regards the equipment referred to in Directives 86/295/EEC,[7] 86/296/EEC[8] and 86/663/EEC,[9] for which these measures shall apply from 1 July 1995.

---

[5] OJ No L 84, 28. 3. 1974, p. 10.
[6] OJ No L 126, 20. 5. 1988, p. 52.
[7] OJ No L 186, 8. 7. 1986, p. 1.
[8] OJ No L 186, 8. 7. 1986, p. 10.
[9] OJ No L 384, 31. 12. 1986, p. 12.

2. Furthermore, Member States shall allow, for the period until 31 December 1994, except as regards the equipment referred to in Directives 86/295/EEC, 86/296/EEC and 86/663/EEC, for which this period shall end on 31 December 1995, the placing on the market and putting into service of machinery in conformity with the national regulations in force in their territory on 31 December 1992.

Directives 86/295/EEC, 86/296/EEC and 86/663/EEC shall not impede implementation of paragraph 1 as from 1 July 1995.

3. Member States shall communicate to the Commission the texts of the provisions of national law which they adopt in the field governed by this Directive.

4. The Commission shall, before 1 January 1994, examine the progress made in the standardization work relating to this Directive and propose any appropriate measures."

6. Annex I is amended as follows:

(a) In section 1.3.7 the following paragraph is added:

"All necessary steps must be taken to prevent accidental blockage of moving parts involved in the work. In cases where, despite the precautions taken, a blockage is likely to occur, specific protection devices or tools, the instruction handbook and possibly a sign on the machinery should be provided by the manufacturer to enable the equipment to be safely unblocked.";

(b) The following section is inserted:

"1.6.5 Cleaning of internal parts

The machinery must be designed and constructed in such a way that it is possible to clean internal parts which have contained dangerous substances or preparations without entering them; any necessary unblocking must also be possible from the outside. If it is absolutely impossible to avoid entering the machinery, the manufacturer must take steps during its construction to allow cleaning to take place with the minimum of danger.";

(c) In section 1.7.0 the following paragraph is added: "Where the health and safety of exposed persons may be endangered by a fault in the operation of unsupervised machinery, the machinery must be equipped to give an appropriate acoustic or light signal as a warning.";

(d) In section 1.7.3 the following paragraphs are added:

"Where a machine part must be handled during use with lifting equipment, its mass must be indicated legibly, indelibly and unambiguously.

The interchangeable equipment referred to in Article 1, section 2, third subparagraph must bear the same information.";

(e) In section 1.7.4 (a) the following indent is added:

"• where necessary, the essential characteristics of tools which may be fitted to the machinery.";

(f) In section 1.7.4 (f) the third paragraph is replaced by the following:

"Where the harmonized standards are not applied, sound levels must be measured using the most appropriate method for the machinery.";

(g) Sections 3 to 5.7 set out in Annex I to this Directive are added.

7. In Annex II (B) the following indents are inserted after the second indent:

"• where appropriate, the name and address of the notified body and the number of the EC type-examination certificate,
• where appropriate, the name and address of the notified body to which the file has been forwarded in accordance with the first indent of Article 8, section 2 (c),
• where appropriate, the name and address of the notified body which has carried out the verification referred to in the second indent of Article 8, section 2 (c),
• where appropriate, a reference to the harmonized standards."

8. In Annex IV, item 12 is replaced by items 12 to 15 set out in Annex II to this Directive.

ARTICLE 2

The following are hereby repealed as from 31 December 1994:

- Articles 2 and 3 of Council Directive 73/361/EEC of 19 November 1973 on the approximation of the laws, regulations and administrative provisions of the Member States relating to the certification and marking of wire ropes, chains and hooks,[10] as lasted amended by Directive 76/434/EEC.[11]

---

[10] OJ No L 335, 5. 12. 1973, p. 51.
[11] OJ No L 122, 8. 5. 1976, p. 20.

- Commission Directive 76/434/EEC of 13 April 1976 adapting to technical progress the Council Directive of 19 November 1973 on the approximation of the laws of the Member States relating to the certification and marking of wire ropes, chains and hooks.

The following are hereby repealed as from 31 December 1995:

- Council Directive 86/295/EEC of 26 May 1986 on the approximation of the laws of the Member States relating to roll-over protective structures (**ROPS**) for certain construction plant.[12]
- Council Directive 86/296/EEC of 26 May 1986 on the approximation of the laws of the Member States relating to falling-object protective structures (**FOPS**) for certain construction plant.[13]
- Council Directive 86/663/EEC of 22 December 1986 on the approximation of the laws of the Member States relating to self-propelled industrial trucks,[14] as last amended by Directive 89/240/EEC.[15]

ARTICLE 3

1. Before 1 January 1992 Member States shall adopt and publish the laws, regulations and administrative provisions necessary in order to comply with this Directive. They shall forthwith inform the Commission thereof.

When Member States adopt these measures, they shall contain a reference to this Directive or shall be accompanied by such reference on the occasion of their official publication. The methods of making such a reference shall be laid down by the Member States.

They shall apply these measures with effect from 1 January 1993.

---

[12] OJ No L 186, 8. 7. 1986, p. 1.
[13] OJ No L 186, 8. 7. 1986, p. 10.
[14] OJ No L 384, 31. 12. 1986, p. 12.
[15] OJ No L 100, 12. 4. 1989, p. 1.

2. Member States shall communicate to the Commission the texts of the provisions of national law which they adopt in the field governed by this Directive.

ARTICLE 4

This Directive is addressed to the Member States.
Done at Luxembourg, 20 June 1991.
For the Council
The President
R. GOEBBELS

ANNEX I

Sections 3 to 5.7 are added to Annex I to Directive 89/392/EEC:

"3. ESSENTIAL HEALTH AND SAFETY REQUIREMENTS TO OFFSET THE PARTICULAR HAZARDS DUE TO THE MOBILITY OF MACHINERY

In addition to the essential health and safety requirements given in the sections 1 and 2, machinery presenting hazards due to mobility must be designed and constructed to meet the requirements below.

Risks due to mobility always exist in the case of machinery which is self-propelled, towed or pushed or carried by other machinery or tractors, is operated in working areas and whose operation requires either mobility while working, be it continuous or semi-continuous movement, between a succession of fixed working positions.

Risks due to mobility may also exist in the case of machinery operated without being moved, but equipped in such a way as to enable it to be moved more easily from one place to another (machinery fitted with wheels, rollers, runners, etc. or placed on gantries, trolleys, etc.).

In order to verify that rotary cultivators and power harrows do not present unacceptable risks to the exposed persons, the manufacturer or his authorized representative established within the Community must, for each type of machinery concerned, perform the appropriate tests or have such tests performed.

3.1 General

3.1.1 Definition

'Driver' means an operator responsible for the movement of machinery. The driver may be transported by the machinery or may be on foot, accompanying the machinery, or may be guiding the machinery by remote control (cables, radio, etc.).

3.1.2 Lighting

If intended by the manufacturer to be used in dark places, self-propelled machinery must be fitted with a lighting device appropriate to the work to be carried out, without prejudice to any other regulations applicable (road traffic regulations, navigation rules, etc.).

3.1.3 Design

Design of machinery to facilitate its handling During the handling of the machine and/or its parts, there must be no possibility of sudden movements or of hazards due to instability as long as the machine and/or its parts are handled in accordance with the manufacturer's instructions.

3.2 Work stations

3.2.1 Driving position

The driving position must be designed with due regard to ergonomic principles. There may be two or more driving positions and, in such cases, each driving position must be provided with all the requisite controls. Where there is more than one driving position, the machinery must be designed so that the use of one of them precludes the use of the others, except in emergency stops. Visibility from the driving position must be such that the driver can in complete safety for himself and the exposed persons, operate the machinery and its tools in their intended conditions of use. Where necessary, appropriate devices must be provided to remedy hazards due to inadequate direct vision.

Machinery must be so designed and constructed that, from the driving position, there can be no risk to the driver and operators on board from inadvertent contact with the wheels or tracks.

The driving position must be designed and constructed so as to avoid any health risk due to exhaust gases and/or lack of oxygen.

The driving position of ride-on drivers must be so designed and constructed that a driver's cab may be fitted as long as there is room. In that case, the cab must incorporate a place for the instructions needed for the driver and/or operators. The driving position must be fitted with an adequate cab where there is a hazard due to a dangerous environment.

Where the machinery is fitted with a cab, this must be designed, constructed and/or equipped to ensure that the driver has good operating conditions and is protected against any hazards that might exist (for instance: inadequate heating and ventilation, inadequate visibility, excessive noise and vibration, falling objects, penetration by objects, rolling over, etc.). The exit must allow rapid evacuation. Moreover, an emergency exit must be provided in a direction which is different from the usual exit.

The materials used for the cab and its fittings must be fire-resistant.

3.2.2 Seating

The driving seat of any machinery must enable the driver to maintain a stable position and be designed with due regard to ergonomic principles.

The seat must be designed to reduce vibrations transmitted to the driver to the lowest level that can be reasonably achieved. The seat mountings must withstand all stresses to which they can be subjected, notably in the event of rollover. Where there is no floor beneath the driver's feet, the driver must have footrests covered with a slip-resistant material.

Where machinery is fitted with provision for a rollover protection structure, the seat must be equipped with a safety belt or equivalent device which keeps the driver in his seat without restricting any movements necessary for driving or any movements caused by the suspension.

3.2.3 Other places

If the conditions of use provide that operators other than the driver are occasionally or regularly transported by the machinery, or work on it, appropriate places must be provided which enable them to be transported or to work on it without risk, particularly the risk of falling.

Where the working conditions so permit, these work places must be equipped with seats.

Should the driving position have to be fitted with a cab, the other places must also be protected against the hazards which justified the protection of the driving position.

3.3 Controls

3.3.1 Control devices

The driver must be able to actuate all control devices required to operate the machinery from the driving position, except for functions which can be safely activated only by using control devices located away from the driving position. This refers in particular to working positions other than the driving position, for which operators other than the driver are responsible or for which the driver has to leave his driving position in order to carry out the manoeuvre in safety.

Where there are pedals they must be so designed, constructed and fitted to allow operation by the driver in safety with the minimum risk of confusion; they must have a slip-resistant surface and be easy to clean.

Where their operation can lead to hazards, notably dangerous movements, the machinery's controls, except for those with preset positions, must return to the neutral position as soon as they are released by the operator.

In the case of wheeled machinery, the steering system must be designed and constructed to reduce the force of sudden movements of the steering wheel or steering lever caused by shocks to the guide wheels. Any control that locks the differential must be so designed and arranged that it allows the differential to be unlocked when the machinery is moving.

The last sentence of section 1.2.2 does not apply to the mobility function.

3.3.2 Starting/moving

Self-propelled machinery with a ride-on driver must be so equipped as to deter unauthorized persons from starting the engine.

Travel movements of self-propelled machinery with a ride-on driver must be possible only if the driver is at the controls.

Where, for operating purposes, machinery must be fitted with devices which exceed its normal clearance zone (e.g. stabilizers, jib, etc.), the driver must be provided with the means of checking easily, before moving the machinery, that such devices are in a particular position which allows safe movement.

This also applies to all other parts which, to allow safe movement, have to be in particular positions, locked if necessary.

Where it is technically and economically feasible, movement of the machinery must depend on safe positioning of the aforementioned parts.

It must not be possible for movement of the machinery to occur while the engine is being started.

3.3.3 Travelling function

Without prejudice to the provisions of road traffic regulations, self-propelled machinery and its trailers must meet the requirements for slowing down, stopping, braking and immobilization so as to ensure safety under all the operating, loading, speed, ground and gradient conditions allowed for by the manufacturer and corresponding to conditions encountered in normal use.

The driver must be able to slow down and stop self-propelled machinery by mean of a main device. Where safety so requires in the event of a failure of the main device, or in the absence of the energy supply to actuate the main device, an emergency device with fully independent and easily accessible controls must be provided for slowing down and stopping.

Where safety so requires, a parking device must be provided to render stationary machinery immobile. This device may be combined with one of the devices referred to in the second paragraph, provided that it is purely mechanical.

Remote-controlled machinery must be designed and constructed to stop automatically if the driver loses control.

Section 1.2.4 does not apply to the travelling function.

3.3.4 Movement of pedestrian-controlled machinery

Movement of pedestrian-controlled self-propelled machinery must be possible only through sustained action on the relevant control by the driver. In particular, it must not be possible for movement to occur while the engine is being started.

The control systems for pedestrian-controlled machinery must be designed to minimize the hazards arising from inadvertent movement of the machine towards the driver. In particular:
 (a) crushing;
 (b) injury from rotating tools.

Also, the speed of normal travel of the machine must be compatible with the pace of a driver on foot.

In the case of machinery on which a rotary tool may be fitted, it must not be possible to actuate that tool when the reversing control is engaged, except where movement of the machinery results from movement of the tool. In the latter case, the reversing speed must be such that it does not endanger the driver

3.3.5 Control circuit failure

A failure in the power supply to the power-assisted steering, where fitted, must not prevent machinery from being steered during the time required to stop it.

3.4 Protection against mechanical hazards

3.4.1 Uncontrolled movements

When a part of a machine has been stopped, any drift away from the stopping position, for whatever reason other than action at the controls, must be such that it is not a hazard to exposed persons.

Machinery must be so designed, constructed and where appropriate placed on its mobile support as to ensure that when moved the uncontrolled oscillations of its centre of gravity do not affect its stability or exert excessive strain on its structure.

3.4.2 Risk of break-up during operation

Parts of machinery rotating at high speed which, despite the measures taken, may break up or disintegrate, must be mounted and guarded in such a way that, in case of breakage, their fragments will be contained or, if that is not possible, cannot be projected towards the driving and/or operation positions.

3.4.3 Rollover

Where, in the case of self-propelled machinery with a ride-on driver and possibly ride-on operators, there is a risk of rolling over, the machinery must be designed for and be fitted with anchorage points allowing it to be equipped with a rollover protective structure (ROPS).

This structure must be such that in case of rolling over if affords the ride-on driver and where appropriate the ride-on operators an adequate deflection-limiting volume (DLV).

In order to verify that the structure complies with the requirement laid down in the second paragraph, the manufacturer or his authorized representative established within the Community must, for each type of structure concerned, perform appropriate tests or have such tests performed.

In addition, the earth-moving machinery listed below with a capacity exceeding 15 kW must be fitted with a rollover protective structure:

- crawler loaders or wheel loaders,
- backhoe loaders,
- crawler tractors or wheel tractors,
- scrapers, self-loading or not,
- graders,
- articulated steer dumpers.

3.4.4 Falling objects

Where, in the case of machinery with a ride-on driver and possibly ride-on operators, there is a risk due to falling objects or material, the machinery should be designed for, and fitted with, if its size allows, anchorage points allowing it to be equipped with a falling-object protective structure (**FOPS**).

This structure must be such that in the case of falling objects or material, it guarantees the ride-on operators an adequate deflection-limiting volume (DLV).

In order to verify that the structure complies with the requirement laid down in the second paragraph, the manufacturer or his authorized representative established within the Community must, for each type of structure concerned, perform appropriate tests or have such tests performed.

3.4.5 Means of access

Handholds and steps must be designed, constructed and arranged in such a way that the operators use them instinctively and do not use the controls for that purpose.

3.4.6 Towing devices

All machinery used to tow or to be towed must be fitted with towing or coupling devices designed, constructed and arranged to ensure easy and safe connection and disconnection, and to prevent accidental disconnection during use.

In so far as the towbar load requires, such machinery must be equipped with a support with a bearing surface suited to the load and the ground.

3.4.7 Transmission of power between self-propelled machinery (or tractor) and recipient machinery

Transmission shafts with universal joints linking self-propelled machinery (or tractor) to the first fixed bearing of recipient machinery must be guarded on the self-propelled machinery side and the recipient machinery side over the whole length of the shaft and associated universal joints.

On the side of the self-propelled machinery (or tractor), the power take-off to which the transmission shaft is attached must be guarded either by a screen fixed to the self-propelled machinery (or tractor) or by any other device offering equivalent protection.

On the towed machinery side, the input shaft must be enclosed in a protective casing fixed to the machinery.

Torque limiters or freewheels may be fitted to universal joint transmissions only on the side adjoining the driven machine. The universal-joint transmission shaft must be marked accordingly.

All towed machinery whose operation requires a transmission shaft to connect it to self-propelled machinery or a tractor must have a system for attaching the transmission shaft so that when the machinery is uncoupled the transmission shaft and its guard are not damaged by contact with the ground or part of the machinery.

The outside parts of the guard must be so designed, constructed and arranged that they cannot turn with the transmission shaft. The guard must cover the transmission shaft to the ends of the inner jaws in the case of simple universal joints and at least to the centre of the outer joint or joints in the case of 'wide-angle' universal joints.

Manufacturers providing means of access to working positions near to the universal joint transmission shaft must ensure that shaft guards as described in the sixth paragraph cannot be used as steps unless designed and constructed for that purpose.

3.4.8 Moving transmission parts

By way of derogation from section 1.3.8.A, in the case of internal combustion engines, removable guards preventing access to the moving parts in the engine compartment need not have locking devices if they have to be opened either by the use of a tool or key or by a control located in the driving position if the latter is in a fully enclosed cab with a lock to prevent unauthorized access.

3.5 Protection against other hazards

3.5.1 Batteries

The battery housing must be constructed and located and the battery installed so as to avoid as far as possible the chance of electrolyte being ejected on to the operator in the event of rollover and/or to avoid the accumulation of vapors in places occupied by operators.

Machinery must be so designed and constructed that the battery can be disconnected with the aid of an easily accessible device provided for that purpose.

3.5.2 Fire

Depending on the hazards anticipated by manufacturer when in use, machinery must, where its size permits:

- either allow easily accessible fire extinguishers to be fitted,
- or be provided with built-in extinguisher systems.

3.5.3 Emissions of dust, gases, etc.

Where such hazards exist, the containment equipment provided for in section 1.5.13 may be replaced by other means, for example precipitation by water spraying.

The second and third paragraphs of section 1.5.13 do not apply where the main function of the machinery is the spraying of products.

3.6 Indications

3.6.1 Signs and warning

Machinery must have means of signalling and/or instruction plates concerning use, adjustment and maintenance, wherever necessary, to ensure the health and safety of exposed persons. They must be chosen, designed and constructed in such a way as to be clearly visible and indelible.

Without prejudice to the requirements to be observed for travelling on the public highway, machinery with a ride-on driver must have the following equipment:

- an acoustic warning device to alert exposed persons,
- a system of light signals relevant to the intended conditions of use such as stop lamps, reversing lamps and rotating beacons. The latter requirement does not apply to machinery intended solely for underground working and having no electrical power.

Remote-controlled machinery which under normal conditions of use exposes persons to the hazards of impact or crushing must be fitted with appropriate means to signal its movements or with means to protect exposed persons against such hazards. The same applies to machinery which involves, when in use, the constant repetition of a forward and backward movement on a single axis where the back of the machine is not directly visible to the driver.

Machinery must be so constructed that the warning and signalling devices cannot all be disabled unintentionally. Where this is essential for safety, such devices must be provided with the means to check that they are in good working order and their failure must be made apparent to the operator.

Where the movement of machinery or its tools is particularly hazardous, signs on the machinery must be provided to warn against approaching the machinery while it is working; the signs must be legible at a sufficient distance to ensure the safety of persons who have to be in the vicinity.

3.6.2 Marking

The minimum requirements set out in 1.7.3 must be supplemented by the following:

- nominal power expressed in kW,
- mass in kg of the most usual configuration and, where appropriate:
- maximum drawbar pull provided for by the manufacturer at the coupling hook, in N,
- maximum vertical load provided for by the manufacturer on the coupling hook, in N.

3.6.3 Instruction handbook

Apart from the minimum requirements set out in 1.7.4, the instruction handbook must contain the following information:

(a) regarding the vibrations emitted by the machinery, either the actual value or a figure calculated from measurements performed on identical machinery:

- the weighted root mean square acceleration value to which the arms are subjected, if it exceeds 2.5 m/s²; should it not exceed 2.5 m/s², this must be mentioned,
- the weighted root mean square acceleration value to which the body (feet or posterior) is subjected, if it exceeds 0.5 m/s²; should it not exceed 0.5 m/s², this must be mentioned.

Where the harmonized standards are not applied, the vibration must be measured using the most appropriate method for the machinery concerned.

The manufacturer must indicate the operating conditions of the machinery during measurement and which methods were used for taking the measurements;

(b) in the case of machinery allowing several uses depending on the equipment used, manufacturers of basic machinery to which interchangeable equipment may be attached and manufacturers of the interchangeable equipment must provide the necessary information to enable the equipment the be fitted and used safely.

4. ESSENTIAL HEALTH AND SAFETY REQUIREMENTS TO OFFSET THE PARTICULAR HAZARDS DUE TO A LIFTING OPERATION

In addition to the essential health and safety requirements given in sections 1, 2 and 3, machinery presenting hazards due to lifting operations—mainly hazards of load falls and collisions or hazards of tipping caused by a lifting operation—must be designed and constructed to meet the requirements below.

Risks due to a lifting operation exist particularly in the case of machinery designed to move a unit load involving a change in level during the movement. The load may consist of objects, materials or goods.

4.1 General remarks

4.1.1 Definitions

(a) lifting accessories:

'lifting accessories' means components or equipment not attached to the machine and placed between the machinery and the load or on the load in order to attach it;

(b) separate lifting accessories:

'separate lifting accessories' means accessories which help to make up or use a slinging device, such as eyehooks, shackles, rings, eyebolts, etc.;

(c) guided load:

'guided load' means the load where the total movement is made along rigid or flexible guides, whose position is determined by fixed points;

(d) working coefficient:

'working coefficient' means the arithmetic ratio between the load guaranteed by the manufacturer up to which a piece of equipment, an accessory or machinery is able to hold it and the maximum working load marked on the equipment, accessory or machinery respectively;

(e) test coefficient:

'test coefficient' means the arithmetic ratio between the load used to carry out the static or dynamic tests on a piece of equipment, an accessory or machinery and the maximum working load marked on the piece of equipment, accessory or machinery;

(f) static test:

'static test' means the test during which the machinery or the lifting accessory is first inspected and the subjected to a force corresponding to the maximum working load multiplied by the appropriate static test coefficient and then re-inspected once the said load has been released to ensure no damage has occurred;

(g) dynamic test:

'dynamic test' means the test during which the machinery is operated in all its possible configurations at maximum working load with account being taken of the dynamic behaviour of the machinery in order to check that the machinery and safety features are functioning properly.

4.1.2 Protection against mechanical hazards

4.1.2.1. Risks due to lack stability

Machinery must be so designed and constructed that the stability required in 1.3.1 is maintained both in service and out of service, including all stages of transportation, assembly and dismantling, during foreseeable component failures and also during the tests carried out in accordance with the instruction handbook.

To that end, the manufacturer or his authorized representative established within the Community must use the appropriate verification methods; in particular, for self-propelled industrial trucks with lift exceeding 1.80 m, the manufacturer or his authorized representative established within the Community must, for each type of industrial truck concerned, perform a platform stability test or similar test, or have such tests performed.

4.1.2.2 Guide rails and rail tracks

Machinery must be provided with devices which act on the guide rails or tracks to prevent derailment.

However, if derailment occurs despite such devices, or if there is a failure of a rail or of a running component, devices must be provided which prevent the equipment, component or load from falling or the machine overturning.

4.1.2.3 Mechanical strength

Machinery, lifting accessories and removable components must be capable of withstanding the stresses to which they are subjected, both in and, where applicable, out of use, under the installation and operating conditions provided for by the manufacturer, and in all relevant configurations, with due regard, where appropriate, to the effects of atmospheric factors and forces exerted by persons. This requirement must also be satisfied during transport, assembly and dismantling. Machinery and lifting accessories must be designed and constructed so as to prevent failure from fatigue or wear, taking due account of their intended use.

The materials used must be chosen on the basis of the working environments provided for by the manufacturer, with special reference to corrosion, abrasion, impacts, cold brittleness and ageing.

The machinery and the lifting accessories must be designed and constructed to withstand the overload in the static tests without permanent deformation or patent defect. The calculation must take account of the values of the static test coefficient chosen to guarantee an adequate level of safety: that coefficient has, as a general rule, the following values:

(a) manually-operated machinery and lifting accessories: 1.5;

(b) other machinery: 1.25.

Machinery must be designed and constructed to undergo, without failure, the dynamic tests carried out using the maximum working load multiplied by the dynamic test coefficient. This dynamic test coefficient is chosen so as to guarantee an adequate level of safety: the coefficient is, as a general rule, equal to 1.1.

The dynamic tests must be performed on machinery ready to be put into service under normal conditions of use. As a general rule, the tests will be performed at the nominal speeds laid down by the manufacturer. Should the control circuit of the machinery allow for a number of simultaneous movements (for example, rotation and displacement of the load), the tests must be carried out under the least favourable conditions, i.e. as a general rule, by combining the movements concerned.

4.1.2.4 Pulleys, drums, chains or ropes

Pulleys, drums and wheels must have a diameter commensurate with the size of rope or chains with which they can be fitted.

Drums and wheels must be so designed, constructed and installed that the ropes or chains with which they are equipped can wind round without falling off.

Ropes used directly for lifting or supporting the load must not include any splicing other than at their ends (splicings are tolerated in installations which are intended from their design to be modified regularly according to needs for use). Complete ropes and their endings have a working coefficient chosen so as to guarantee an adequate level of safety; as a general rule, this coefficient is equal to five.

Lifting chains have a working coefficient chosen so as to guarantee an adequate level of safety; as a general rule, this coefficient is equal to four.

In order to verify that an adequate working coefficient has been attained, the manufacturer or his authorized representative established within the Community must, for each type of chain and rope used directly for lifting the load, and for the rope ends, perform the appropriate tests or have such tests performed.

4.1.2.5 Separate lifting accessories

Lifting accessories must be sized with due regard to fatigue and ageing processes for a number of operating cycles consistent with their expected life-span as specified in the operating conditions for a given application.

Moreover:

(a) the working coefficient of the metallic rope/rope-end combination is chosen so as to guarantee an adequate level of safety; this coefficient is, as a general rule, equal to five. Ropes must not comprise any splices or loops other than at their ends;

(b) where chains with welded links are used, they must be of the short-link type. The working coefficient of chains of any type is chosen so as to guarantee an adequate level of safety; this coefficient is, as a general rule, equal to four;

(c) the working coefficient for textile ropes or slings is dependent on the material, method of manufacture, dimensions and use. This coefficient is chosen so as to guarantee an adequate level of safety; it is, as a general rule, equal to seven, provided the materials used are shown to be of very good quality and the method of manufacture is appropriate to the intended use. Should this not be the case, the coefficient is, as a general rule, set at a higher level in order to secure an equivalent level of safety.

Textile ropes and slings must not include any knots, connections or splicing other than at the ends of the sling, except in the case of an endless sling;

(d) all metallic components making up, or used with, a sling must have a working coefficient chosen so as to guarantee an adequate level of safety; this coefficient is, as a general rule, equal to four;

(e) the maximum working capacity of a multi-legged sling is determined on the basis of the safety coefficient of the weakest leg, the number of legs and a reduction factor which depends on the slinging configuration;

(f) in order to verify that an adequate working coefficient has been attained, the manufacturer or his authorized representative established within the Community must, for each type of component referred to in (a), (b), (c) and (d) perform the appropriate tests or have such tests performed.

4.1.2.6 Control of movements

Devices for controlling movements must act in such a way that the machinery on which they are installed is kept safe:

(a) machinery must be so designed or fitted with devices that the amplitude of movement of its components is kept within the specified limits. The operation of such devices must, where appropriate, be preceded by a warning;

(b) where several fixed or rail-mounted machines can be manoeuvred simultaneously in the same place, with risks of collision, such machines must be so designed and constructed as to make it possible to fit systems enabling these risks to be avoided,

(c) the mechanisms of machinery must be so designed and constructed that the loads cannot creep dangerously of fall freely and unexpectedly, even in the event of partial or total failure of the power supply or when the operator stops operating the machine;

(d) it must not be possible, under normal operating conditions, to lower the load solely by friction brake, except in the case of machinery, whose function requires it to operate in that way;

(e) holding devices must be so designed and constructed that inadvertent dropping of the loads is avoided.

4.1.2.7 Handling of loads

The driving position of machinery must be located in such a way as to ensure that widest possible view of trajectories of the moving parts, in order to avoid possible collisions with persons or equipment or other machinery which might be manoeuvring at the same time and liable to constitute a hazard.

Machinery with guided loads fixed in one place must be designed and constructed so as to prevent exposed persons from being hit by the load or the counter-weights.

4.1.2.8 Lightning

Machinery in need of protection against the effects of lightning while being used must be fitted with a system for conducting the resultant electrical charges to earth.

4.2 Special requirements for machinery whose power source is other than manual effort

4.2.1 Controls

4.2.1.1 Driving position

The requirements laid down in section 3.2.1 also apply to non-mobile machinery.

4.2.1.2 Seating

The requirements laid down in section 3.2.2, first and second paragraphs, and those laid down in section 3.2.3 also apply to non-mobile machinery.

4.1.2.3 Control devices

The devices controlling movements of the machinery or its equipment must return to their neutral position as soon as they are released by the operator. However, for partial or complete movements in which there is no risk of the load or the machinery colliding, the said devices may be replaced by controls authorizing automatic stops at preselected levels without holding a hold-to-run control device.

4.1.2.4 Loading control

Machinery with a maximum working load of not less than 1000 kilograms or an overturning moment of not less than 40,000 N · m must be fitted with devices to warn the driver and prevent dangerous movements of the load in the event of:

- overloading the machinery
- either as a result of maximum working loads being exceeded, or
- as a result of the moments due to the loads being exceeded,
- the moments conducive to overturning being exceeded as a result of the load being lifted.

4.2.2 Installation guided by cables

Cable carriers, tractors or tractor carriers must be held by counterweights or by a device allowing permanent control of the tension.

4.2.3 Risks to exposed persons. Means of access to driving position and intervention points

Machinery with guided loads and machinery whose load supports follow a clearly defined path must be equipped with devices to prevent any risks to exposed persons.

4.2.4 Fitness for purpose

When machinery is placed on the market or is first put into service, the manufacturer or his authorized representative established within the Community must ensure, by taking appropriate measures or having them taken, that lifting accessories and machinery which are ready for use—whether manually or power-operated—can fulfil their specified functions safely. The said measures must take into account the static and dynamic aspects of the machinery.

Where the machinery cannot be assembled in the manufacturer's premises, or in the premises of his authorized representative established within the Community, appropriate measures must be taken at the place of use. Otherwise, the measures may be taken either in the manufacturer's premises or at the place of use.

4.3 Marking

4.3.1 Chains and ropes

Each length of lifting chain, rope or webbing not forming part of an assembly must bear a mark or, where this is not possible, a plate or irremovable ring bearing the name and address of the manufacturer or his authorized representative established in the Community and the identifying reference of the relevant certificate.

The certificate should show the information required by the harmonized standards or, should those not exist, at least the following information:

- the name of the manufacturer or his authorized representative established within the Community,
- the address within the Community of the manufacturer or his authorized representative, as appropriate,
- a description of the chain or rope which includes:
  - its nominal size,
  - its construction,
  - the material from which it is made, and
  - any special metallurgical treatment applied to the material,
- if tested, the standard used,
- a maximum load to which the chain or rope should be subjected in service. A range of values may be given for specified applications.

4.3.2 Lifting accessories

All lifting accessories must show the following particulars:

- identification of the manufacturer,
- identification of the material (e.g. international classification) where this information is needed for dimensional compatibility,
- identification of the maximum working load,
- EC mark.

In the case of accessories including components such as cables or ropes, on which marking is physically impossible, the particulars referred to in the first paragraph must be displayed on a plate or by some other means and securely affixed to the accessory.

The particulars must be legible and located in a place where they are not liable to disappear as a result of machining, wear, etc., or jeopardize the strength of the accessory.

4.3.3 Machinery

In addition to the minimum information provided for in 1.7.3, each machine must bear, legibly and indelibly, information concerning the nominal load:

(i) displayed in uncoded form and prominently on the equipment in the case of machinery which has only one possible value;

(ii) where the nominal load depends on the configuration of the machine, each driving position must be provided with a load plate indicating, preferably in diagrammatic form or by means of tables, the nominal loads for each configuration.

Machinery equipped with a load support which allows access to persons and involves a risk of falling must bear a clear and indelible warning prohibiting the lifting of persons. This warning must be visible at each place where access is possible.

4.4 Instruction handbook

4.4.1 Lifting accessories

Each lifting accessory or each commercially indivisible batch of lifting accessories must be accompanied with an instruction handbook setting out at least the following particulars:

- normal conditions of use,
- instructions for use, assembly and maintenance,
- the limits of use (particularly for the accessories which cannot comply with 4.1.2.6 (e)).

4.4.2 Machinery

In addition to section 1.7.4, the instruction handbook must include the following information:

(a) the technical characteristics of the machinery, and in particular:

- where appropriate, a copy of the load table described in section 4.3.3 (ii),
- the reactions at the supports or anchors and characteristics of the tracks,
- where appropriate, the definition and the means of installation of the ballast;

(b) the contents of the logbook, if the latter is not supplied with the machinery;

(c) advice for use, particularly to offset the lack of direct sight of the load by the operator;

(d) the necessary instructions for performing the tests before first putting into service machinery which is not assembled on the manufacturer's premises in the form in which it is to be used.

5. ESSENTIAL SAFETY AND HEALTH REQUIREMENTS FOR MACHINERY INTENDED SOLELY FOR UNDERGROUND WORK

In addition to the essential safety and health requirements provided for in sections 1, 2, 3 and 4, machinery intended solely for underground work must be designed and constructed to meet the requirements below.

5.1 Risks due to lack of stability. Powered roof supports must be so designed and constructed as to maintain a given direction when moving and not slip before and while they come under load and after the load has been removed. They must be equipped with anchorages for the top plates of the individual hydraulic props.

5.2 Movement. Powered roof supports must allow for unhindered movement of exposed persons.

5.3 Lighting. The requirements laid down in the third paragraph of section 1.1.4 do not apply.

5.4 Control devices. The accelerator and brake controls for the movement of machinery running on rails must be manual. The deadman's control may be foot-operated, however.

The control devices of powered roof supports must be designed and laid out so that, during displacement operations, operators are sheltered by a support in place. The control devices must be protected against any accidental release.

5.5 Stopping. Self-propelled machinery running on rails for use in underground work must be equipped with a deadman's control acting on the circuit controlling the movement of the machinery.

5.6 Fire. The second indent of 3.5.2 is mandatory in respect of machinery which comprises highly flammable parts.

The braking system of machinery meant for use in underground working must be designed and constructed so as not produce sparks or cause fires.

Machinery with heat engines for use in underground working must be fitted only with internal combustion engines using fuel with a low vaporizing pressure and which exclude any spark of electrical origin.

5.7 Emissions of dust, gases etc. Exhaust gases from internal combustion engines must not be discharged upwards."

ANNEX II

Items 12 to 15 are added to Annex IV to Directive 89/392/EEC:
"12. Machinery for underground working of the following types:

- machinery on rails: locomotives and brake-vans,
- hydraulic-powered roof supports,

- internal combustion engines to be fitted to machinery for underground working.

13. Manually-loaded trucks for the collection of household refuse incorporating a compression mechanism.
14. Guards and detachable transmission shafts with universal joints as described in section 3.4.7.
15. Vehicles servicing lifts."

DIRECTIVE 93/44/EEC

COUNCIL DIRECTIVE 93/44/EEC of 14 June 1993 amending Directive 89/392/EEC on the approximation of the laws of the Member States relating to machinery

THE COUNCIL OF THE EUROPEAN COMMUNITIES,

Having regard to the Treaty establishing the European Economic Community, and in particular Article 100a thereof,

Having regard to the proposal from the Commission.[1]

In cooperation with the European Parliament.[2]

Having regard to the opinion of the Economic and Social Committee.[3]

Whereas lifting persons entails specific hazards for the persons lifted; whereas those hazards are not covered by the essential health and safety requirements laid down by Council Directive 89/392/EEC of 14 June 1989 on the approximation of the laws of the Member States relating to machinery.[4]

Whereas, for the type of machinery in question, there is no reason to provide for different conformity assessment modules from those initially provided for in Directive 89/392/EEC for machinery in general;

Whereas additional essential health and safety requirements to cover the risks run by lifted persons can be laid down by means of an amendment to Directive 89/392/EEC; whereas this amendment can be used to advantage to correct certain imperfections in the said Directive;

Whereas it is also necessary to deal with safety components which are placed on the market separately and the safety function of which is declared by the manufacturer or his authorized representative established in the Community;

---

[1] OJ No C25, 1. 2. 1992, p. 8; and OJ No C 252, 29. 9. 1992, p. 3.
[2] OJ No C 241, 21.9. 1992, p. 107; and OJ No C 72, 15. 3. 1993, p. 70.
[3] OJ No C 223, 31. 8. 1992, p. 60.
[4] OJ No L 183, 29. 6. 1989, p. 9. Directive as amended by Directive 91/368/EEC (OJ No L 198, 22. 7.1991, p. 16).

Whereas the dates of implementation laid down in this Directive shall not alter the dates of implementation of Directive 89/392/EEC and Directive 91/368/EEC, which amended it,

HAS ADOPTED THIS DIRECTIVE:

ARTICLE 1

Directive 89/392/EEC is hereby amended as follows:

1. Article 1 shall be amended as follows:

(a) the following subparagraph shall be added to paragraph 1:

"It shall also apply to safety components placed on the market separately.";

(b) the following subparagraph shall be added to paragraph 2:

"For the purposes of this Directive, safety components means a component, provided that it is not interchangeable equipment, which the manufacturer or his authorized representative established in the Community places on the market to fulfil a safety function when in use and the failure or malfunctioning of which endangers the safety or health of exposed persons.";

(c) paragraph 3 shall be amended as follows:

(i) the following indent shall be deleted:

"• lifting equipment designed and constructed for raising and/or moving persons with or without loads, except for industrial trucks with elevating operation position,";

(ii) the following indent:

"• cableways for the public or private transposition of persons," shall be replaced by:

"• cableways, including funicular railways, for the public or private transportation of persons,";

(iii) the following indents shall be added:

"• lifts which permanently serve specific levels of buildings and constructions, having a car moving between guides which are rigid

and inclined at a angle of more than 15 degrees to the horizontal and designed for the transport of:
- persons,
- persons and goods,
- goods alone if the car is accessible, that is to say, a person may enter it without difficulty, and fitted with controls situated inside the car or within reach of a person inside,
- means of transport of persons using rack and pinion rail mounted vehicles,
- mine winding gear,
- theatre elevators,
- construction site hoists intended for lifting persons or persons and goods.";

(d) paragraph 4 shall be replaced by the following:
"4. Where, for machinery or safety components, the risks referred to in this Directive are wholly or partly covered by specific Community directives, this Directive shall not apply, or shall cease to apply, in the case of such machinery or safety components and of such risks on the implementation of these specific Directives."

2. Article 2 shall be replaced by the following:

"ARTICLE 2

1. Member States shall take all appropriate measures to ensure that machinery or safety components covered by this Directive may be replaced on the market and put into service only if they do not endanger the health or safety of persons and, where appropriate, domestic animals or property, when properly installed and maintained and used for their intended purpose.

2. This Directive shall not affect Member States' entitlement to lay down, in due observance of the Treaty, such requirements as they may deem necessary to ensure that persons and in particular workers are protected when using the machinery or safety components in question, provided that this does not mean that the machinery or safety components are modified in a way not specified in the Directive.

3. At trade fairs, exhibitions, demonstrations, etc., Member States shall not prevent the showing of machinery or safety components which do not conform to the provisions of this Directive, provided that a visible sign clearly indicates that such machinery or safety components do not conform and that they are not for sale until they have been brought into conformity by the manufacturer or his authorized representative established in the Community. During demonstrations, adequate safety measures shall be taken to ensure the protection of persons.";

3. Article 3 shall be replaced by the following:

"ARTICLE 3

Machinery and safety components covered by this Directive shall satisfy the essential health and safety requirements set out in Annex I.";

4. Article 4 shall be amended as follows:

(a) paragraph 1 shall be replaced by the following:

"1. Member States shall not prohibit, restrict or impede the placing on the market and putting into service in their territory of machinery and safety components which comply with this Directive.";

(b) the following paragraph shall be added:

"3. Member States may not prohibit, restrict or impede the placing on the market of safety components as defined in Article 1, section 2, where they are accompanied by an EC declaration of conformity by the manufacturer or his authorized representative established in the Community as referred to in Annex II, point C.";

5. Article 5, section 1 and section 2 shall be replaced by the following:

"1. Member States shall regard the following as conforming to all the provisions of this Directive, including the procedures for checking the conformity provided for in Chapter II:

- machinery bearing the CE marking and accompanied by the EC declaration of conformity referred to in Annex II.A.,
- safety components accompanied by the EC declaration of conformity referred to in Annex II.C.

In the absence of harmonized standards, Member States shall take any steps they deem necessary to bring to the attention of the parties concerned the existing national technical standards and specifications which are regarded as important or relevant to the proper implementation of the essential safety and health requirements in Annex I.

2. Where a national standard transposing a harmonized standard, the reference for which has been published in the Official Journal of the European Communities, covers one or more of the essential safety requirements, machinery or safety components constructed in accordance with this standard shall be presumed to comply with the relevant essential requirements.

Member States shall publish the references of national standards transposing harmonized standards.";

6. Article 7 shall be amended as follows:

(a) paragraph 1 shall be replaced by the following:
"1. Where a Member State ascertains that

- machinery bearing the CE marking,

or

- safety components accompanied by the EC declaration of conformity, used in accordance with their intended purpose are liable to endanger the safety of persons, and, where appropriate, domestic animals or property, it shall take all appropriate measures to withdraw such machinery or safety components from the market, to prohibit the placing on the market, putting into service or use thereof, or to restrict free movement thereof.

The Member States shall immediately inform the Commission of any such measure, indicating the reason for its decision and, in particular, whether non-conformity is due to:

(a) failure to satisfy the essential requirements referred to in Article 3;

(b) incorrect application of the standards referred to in Article 5, section 2;

(c) shortcomings in the standards referred to in Article 5, section 2, themselves.";

(b) paragraph 3 shall be replaced by the following:

"3. Where:

- machinery which does not comply bears the CE marking,
- a safety component which does not comply is accompanied by an EC declaration of conformity,

the competent Member State shall take appropriate action against whomsoever has affixed the marking or drawn up the declaration and shall so inform the Commission and other Member States.";

7. Article 8 shall be amended as follows:
(a) paragraph 1 shall be replaced by the following:
"1. The manufacturer or his authorized representative established in the Community must, in order to certify that machinery and safety components are in conformity with this Directive, draw up for all machinery or safety component manufactured an EC declaration of conformity based on the model given in Annex II, A or C as appropriate.

In addition, for machinery alone, the manufacturer or his authorized representatives established in the Community must affix to the machine the CE marking referred to in Article 10.";

(b) the following paragraph shall be inserted:
"4a. Safety components shall be subject to the certification procedures applicable to machinery pursuant to paragraphs 2, 3 and 4. Furthermore, during EC type-examination, the notified body shall verify the suitability of the safety component for fulfilling the safety functions declared by the manufacturer.";

(c) paragraph 6 shall be replaced by the following:
"6. Where neither the manufacturer nor his authorized representative established in the Community fulfils the obligations of the preceding paragraphs, these obligations shall fall to any person placing the machinery or safety component on the market in the Community. The same obligations shall apply to any person assembling machinery or parts thereof or safety components of various origins or constructing machinery or safety components for his own use.";

8. Article 9, section 1, shall be replaced by the following:
"1. Each Member State shall notify the Commission and the other Member States of the bodies responsible for carrying out the certification procedures referred to in Article 8. The Commission shall publish a list of those bodies in the Official Journal of the European Communities for information and shall ensure that the list is kept up to date.";

9. Article 11 shall be replaced by the following:

"ARTICLE 11

Any decision taken pursuant to this Directive which restricts the placing on the market and putting into service of machinery or a safety component shall state the exact grounds on which it is based. Such a decision shall be notified as soon as possible to the party concerned, who shall at the same time be informed of the legal remedies available to him under the laws in force in the Member State concerned and of the time limits to which such remedies are subject.";

10. Annex I shall be amended as follows:
(a) the title shall be replaced by the following:

"ESSENTIAL HEALTH AND SAFETY REQUIREMENTS RELATING TO THE DESIGN AND CONSTRUCTION OF MACHINERY AND SAFETY COMPONENTS";

(b) the following shall be added after this title:
"For the purposes of this Annex, 'machinery' means either 'machinery' or 'safety component' as defined in Article 1, section 2.";

(c) the preliminary observations shall be supplemented by the following:

"3. The essential health and safety requirements have been grouped according to the hazards which they cover.

Machinery presents a series of hazards which may be indicated under more than one heading in this Annex.

The manufacturer is under an obligation to assess the hazards in order to identify all those which apply to his machine; he must then design and construct it taking account of his assessment.";

(d) in Section 1.2.4, the last paragraph concerning the emergency stop shall be replaced by the following:

"Once active operation of the emergency stop control has ceased following a stop command, that command must be sustained by engagement of the emergency stop device until that engagement is specifically overridden; it must not be possible to engage the device without triggering a stop command; it must be possible to disengage the device only by an appropriate operation, and disengaging the device must not restart the machinery but only permit restarting.";

(e) the following sections shall be added:

"1.5.14 Risk of being trapped in a machine

Machinery must be designed, constructed or fitted with a means of preventing an exposed person from being enclosed within it or, if that is impossible, with a means of summoning help.

1.5.15 Risk of slipping, tripping or falling

Parts of the machinery where persons are liable to move about or stand must be designed and constructed to prevent persons slipping, tripping or falling on or off these parts.";

(f) the second paragraph of Section 1.6.2 shall be deleted;

(g) the first indent of Section 1.7.4 (a) shall be replaced by the following:

> "• a repeat of the information with which the machinery is marked, except the serial number (see 1.7.3) together with any appropriate additional information to facilitate maintenance (e.g. addresses of the importer, repairers etc.),";

(h) Section 1.7.4 (b) shall be replaced by the following:
"The constructions must be drawn up in one of the Community languages by the manufacturer or his authorized representative established in the Community. On being put into service, all machinery must be accompanied by a translation of the instructions in the language or languages of the country in which the machinery is to be used and by the instructions in the original language. This translation must be done either by the manufacturer or his authorized representative established in the Community or by the person introducing the machinery into the language area in question. By way of derogation from this requirement, the maintenance instructions for use by specialized personnel employed

by the manufacturer or his authorized representative established in the Community may be drawn up in only one of the Community languages understood by that personnel.";

(i) Section 1.7.4 (d) shall be replaced by the following:

"Any literature describing the machinery must not contradict the instructions as regards safety aspects. The technical documentation describing the machinery must give information regarding the airborne noise emissions referred to in (f) and, in the case of hand-held and/or hand-guided machinery, information regarding vibration as referred to in 2.2.";

(j) the title of Section 2 shall be replaced by the following:

"ESSENTIAL HEALTH AND SAFETY REQUIREMENTS FOR CERTAIN CATEGORIES OF MACHINERY";

(k) in Sections 2.1, 2.2 and 2.3, the following words shall be deleted:
"In addition to the essential health and safety requirements set out in 1 above,";

(l) in Section 3, the first paragraph shall be replaced by the following:
"Machinery presenting hazards due to mobility must be designed and constructed to meet the requirements set out below.";

(m) in Section 4, the first paragraph shall be replaced by the following:
"Machinery presenting hazards due to lifting operations—mainly hazards of load falls and collisions or hazards of tipping caused by a lifting operation—must be designed and constructed to meet the requirements set out below.";

(n) the following paragraph shall be added to Section 4.2.3:
"Machinery serving specific levels at which operators can gain access to the load platform in order to stack or secure the load must be designed and constructed to prevent uncontrolled movement of the load platform, in particular while being loaded or unloaded.";

(o) the title of Section 5 shall be replaced by the following:

"ESSENTIAL HEALTH AND SAFETY REQUIREMENTS FOR MACHINERY INTENDED FOR UNDERGROUND WORK";

(p) the first paragraph of Section 5 shall be replaced by the following: "Machinery intended for underground work must be designed and constructed to meet the requirements set out below.";

(q) Section 6 in the Annex to this Directive shall be added;

11. Annex II shall be amended as follows:

(a) the title of point A shall be replaced by the following:

"Contents of the EC declaration of conformity for machinery.[1]"

(b) the first footnote shall be replaced by the following:

"(1) This declaration must be drawn up in the same language as the original instructions (see Annex I, Section 1.7.4 (b)) and must be either typewritten or handwritten in block capitals. It must be accompanied by a translation in one of the official languages of the country in which the machinery is to be used. This translation must be done in accordance with the same conditions as for the translation of the instructions.";

(c) the following point shall be added:

"C. Contents of the EC declaration of conformity for safety components placed on the market separately.[1]

The EC declaration of conformity must contain the following particulars:

- name and address of the manufacturer or his authorized representative established in the Community,[2]
- description of the safety component,[4]
- safety function fulfilled by the safety component, if not obvious from the description,
- where appropriate, the name and address of the notified body and the number of the EC type-examination certificate,
- where appropriate, the name and address of the notified body to which the file was forwarded in accordance with the first indent of Article 8, section 2 (c),
- where appropriate, the name and address of the notified body which carried out the verification referred to in the second indent of Article 8, section 2 (c),
- where appropriate, a reference to the harmonized standards,
- where appropriate, the national technical standards and specifications used,

- identification of the person empowered to sign on behalf of the manufacturer or his authorized representative established in the Community.";

(d) the following footnote 4 shall be added:
"(4) Description of the safety component (make, type, serial number, if any, etc.).";

12. Annex IV shall be amended as follows:
(a) the title shall be replaced by the following:

TYPES OF MACHINERY AND SAFETY COMPONENTS FOR WHICH THE PROCEDURE REFERRED TO IN ARTICLE 8, SECTION 2 (B) AND (C) MUST BE APPLIED";

(b) the following subtitle shall be inserted after the title:
"A. Machinery";
(c) Section 1 shall be replaced by the following:
"1. Circular saws (single or multi-blade) for working with wood and analogous materials or for working with meat and analogous materials.

1.1 Sawing machines with fixed tool during operation, having a fixed bed with manual feed of the workpiece or with a demountable power feed.

1.2 Sawing machines with fixed tool during operation, having a manually operated reciprocating saw-bench or carriage.

1.3 Sawing machines with fixed tool during operation, having a built-in mechanical feed device for the workpieces, with manual loading and/or unloading.

1.4 Sawing machines with movable tool during operation, with a mechanical feed device and manual loading and/or unloading.";

(d) Section 4 shall be replaced by the following:
"4. Band-saws with a fixed or mobile bed and band-saws with a mobile carriage, with manual loading and/or unloading, for working with wood and analogous materials or for working with meat and analogous materials.";

(e) Section 5 shall be replaced by the following:
"5. Combined machines of the types referred to in 1 to 4 and 7 for working with wood and analogous materials.";

(f) Section 7 shall be replaced by the following:

"7. Hand-fed vertical spindle moulding machines for working with wood and analogous materials.";

(g) the following shall be added to Section A:

"16. Devices for the lifting of persons involving a risk of falling from a vertical height of more than three metres.

17. Machines for the manufacture of pyrotechnics.";

(h) the following section shall be added:

"B. Safety components

1. Electro-sensitive devices designed specifically to detect persons in order to ensure their safety (non-material barriers, sensor mats, electro-magnetic detectors, etc.).

2. Logic units which ensure the safety functions of bi-manual controls.

3. Automatic movable screens to protect the presses referred to in 9, 10 and 11.

4. Roll-over protection structures (ROPS).

5. Fallings-object protective structures (FOPS).";

13. the following shall be added after the title in Annex V:

"For the purposes of this Annex, 'machinery' means either 'machinery' or 'safety component' as defined in Article 1, section 2.";

14. the following shall be added after the title in Annex VI:

"For the purposes of this Annex, 'machinery' means either 'machinery' or 'safety component' as defined in Article 1, section 2.";

15. the following shall be added after the title in Annex VII:

"For the purposes of this Annex, 'machinery' means either 'machinery' or 'safety component' as defined in Article 1, section 2."

ARTICLE 2

1. Before 1 July 1994 Member States shall adopt and publish the laws, regulations and administrative provisions necessary in order to comply with this Directive. They shall forthwith inform the Commission thereof.

When Member States adopt these measures, they shall contain a reference to this Directive or shall be accompanied by such reference on the occasion of their official publication. The methods of making such a reference shall be laid down by the Member States.

Member States shall apply these provisions with effect from 1 January 1995.

2. By way of derogation from the third subparagraph of paragraph 1, Member States shall apply the laws, regulations and administrative provisions necessary to comply with the provisions listed below as from 1 July 1994:

- Article 1, section 10, excluding (a), (b) and (q),
- Article 1, section 11 (a) and (b),
- Article 1, section 12 (c), (d), (e) and (f).

3. However, until 31 December 1996 Member States may allow the placing on the market and putting into service of safety components and of machinery for the lifting or moving of persons which conform with the national provisions in force in their territories as at the date of adoption of this Directive.

4. Member States shall communicate to the Commission the text of the provisions of national law which they adopt in the field covered by this Directive.

ARTICLE 3

This Directive is addressed to the Member States.
Done at Luxembourg, 14 June 1993.
For the Council
The President
J. TROEJBORG

ANNEX

"6. ESSENTIAL HEALTH AND SAFETY REQUIREMENTS TO OFFSET THE PARTICULAR HAZARDS DUE TO THE LIFTING OR MOVING OF PERSONS.

Machinery presenting hazards due to the lifting or moving of persons must be designed and constructed to meet the requirements set out below.

6.1 General

6.1.1 Definition. For the purposes of this Chapter, 'carrier' means the device by which persons are supported in order to be lifted, lowered or moved.

6.1.2 Mechanical strength. The working coefficients defined in heading 4 are inadequate for machinery intended for the lifting or moving of persons and must, as a general rule, be doubled. The floor of the carrier must be designed and constructed to offer the space and strength corresponding to the maximum number of persons and the maximum working load set by the manufacturer.

6.1.3 Loading control for types of device moved by power other than human strength. The requirements of 4.2.1.4 apply regardless of the maximum working load figure. This requirement does not apply to machinery in respect of which the manufacturer can demonstrate that there is no risk of overloading and/or overturning.

6.2 Controls

6.2.1 Where safety requirements do not impose other solutions:
The carrier must, as a general rule, be designed and constructed so that persons inside have means of controlling movements upwards and downwards and, if appropriate, of moving the carrier horizontally in relation to the machinery.

In operation, those controls must override the other devices controlling the same movement, with the exception of the emergency stop devices.

The controls for these movement must be of the maintained command type, except in the case of machinery serving specific levels.

6.2.2 If machinery for the lifting or moving of persons can be moved with the carrier in a position other than the rest position, it must be designed and constructed so that the person or persons in the carrier have the means of preventing hazards produced by the movement of the machinery.

6.2.3 Machinery for the lifting or moving of persons must be designed, constructed or equipped so that excess speeds of the carrier do not cause hazards.

6.3 Risks of persons falling from the carrier

6.3.1 If the measures referred to in 1.1.15 are not adequate, carriers must be fitted with a sufficient number of anchorage points for the number of persons possibly using the carrier, strong enough for the attachment of personal protective equipment against the danger of falling.

6.3.2 Any trapdoors in floors or ceilings or side doors must open in a direction which obviates any risk of falling should they open unexpectedly.

6.3.3 Machinery for lifting or moving must be designed and constructed to ensure that the floor of the carrier does not tilt to an extend which creates a risk of the occupants falling, including when moving.

The floor of the carrier must be slip-resistant.

6.4 Risks of the carrier falling or overturning

6.4.1 Machinery for the lifting or moving of persons must be designed and constructed to prevent the carrier falling or overturning.

6.4.2 Acceleration and braking of the carrier or carrying vehicle, under the control of the operator or triggered by a safety device and under the maximum load and speed conditions laid down by the manufacturer, must not cause any danger to exposed persons.

6.5 Markings

Where necessary to ensure safety, the carrier must bear the relevant essential information."

Appendix II

List of Standards

European standardization body	Reference	Title of the harmonized standards	Year of ratification
CEN	EN 115	Safety rules for the constuction and installation of escalators and passenger conveyors	1995
CEN	EN 201	Rubber and plastic machines—Injection moulding machines—Safety	1997
CEN	EN 289	Rubber and plastic machinery—Compression and transfer moulding presses—Safety requirements for the design	1993
CEN	EN 292-1	Safety of machinery—Basic concepts, general principles for design—Part 1: Basic terminology, methodology	1991
CEN	EN 292-2	Safety of machinery—Basic concepts, general principles for design—Part 2: Technical principles and specifications	1991
CEN	EN 292-2/A1	Safety of machinery—Basic concepts, general principles for design—Part 2: Technical principles and specifications	1995
CEN	EN 294	Safety of machinery—Safety distance to prevent danger zones being reached by the upper limbs	1992
CEN	EN 349	Safety of machinery—Minimum gaps to avoid crushing of parts of the human body	1993
CEN	EN 415-4	Safety of packaging machines—Part 4: Palletizers and depalletizers	1997
CEN	EN 418	Safety of machinery—Emergency stop equipment, functional aspects—Principles for design	1992
CEN	EN 422	Rubber and plastic machines—Safety—Blow moulding machines intended for the production of hollow articles—Requirements for the design and construction	1995
CEN	EN 457	Safety of machinery—Auditory danger signals—General requirements, design and testing (ISO 7731: 1986, modified)	1992
CEN	EN 474-1	Earth—moving machinery—Safety—Part 1: General requirements	1994

CEN	EN 474-2	Earth—moving machinery—Safety—Part 2: Requirements for tractor—dozers	1996
CEN	EN 474-3	Earth—moving machinery—Safety—Part 3: Requirements for loaders	1996
CEN	EN 474-4	Earth—moving machinery—Safety—Part 4: Requirements for backhoe loaders	1996
CEN	EN 474-5	Earth—moving machinery—Safety—Part 5: Requirements for hydraulic excavators	1996
CEN	EN 474-6	Earth—moving machinery—Safety—Part 6: Requirements for dumpers	1996
CEN	EN 500-1	Mobile road construction machinery—Safety—Part 1: Common requirements	1995
CEN	EN 500-2	Mobile road construction machinery—Safety—Part 2: Specific requirements for road—milling machines	1995
CEN	EN 500-3	Mobile road construction machinery—Safety—Part 3: Specific requirements for soil stabilization machines	1995
CEN	EN 500-4	Mobile road construction machinery—Safety—Part 4: Specific requirements for compaction machines	1995
CEN	EN 500-5	Mobile road construction machinery—Safety—Part 5: Specific requirements for joint cutters	1995
CEN	EN 528	Rail dependent storage and retrieval equipment—Safety	1996
CEN	EN 547-1	Safety of machinery—Human body measurements—Part 1: Principles for determining the dimensions required for openings for the whole body access into machinery	1996
CEN	EN 547-2	Safety of machinery—Human body measurements—Part 2: Principles for determining the dimensions required for access openings	1996
CEN	EN 547-3	Safety of machinery—Human body measurements—Part 3: Anthropometric data	1996
CEN	EN 563	Safety of machinery—Temperatures of touchable surfaces—Ergonomics data to establish temperature limit values for hot surfaces	1994

Appendix II: List of Standards

CEN	EN 574	Safety of machinery—Two—hand control devices—Functional aspects—Principles for design	1996
CEN	EN 608	Agricultural and forestry machinery—Portable chainsaws—Safety	1994
CEN	EN 614-1	Safety of machinery—Egonomic design principles—Part 1: Terminology and general principles	1995
CEN	EN 626-1	Safety of machinery—Reduction of risks to health from hazardous substances emitted by machinery—Part 1: Principles and specifications for machinery manufacturers	1994
CEN	EN 626-2	Safety of machinery—Reduction of risks to health from hazardous substances emitted by machinery—Part 2: Methodology leading to verification procedures	1996
CEN	EN 627	Specification for data—logging and monitoring of lifts, escalators and passenger conveyors	1995
CEN	EN 632	Agricultural machinery—Combine harvesters and forage harvesters—Safety	1995
CEN	EN 690	Agricultural machinery—Manure spreaders—Safety	1994
CEN	EN 703	Agricultural machinery—Silage cutters—Safety	1995
CEN	EN 706	Safety requirements for agricultural and forestry machinery—Vine shoot tipping machines	1996
CEN	EN 708	Agricultural machinery—Soil working machines with powered tools—Safety	1996
CEN	EN 709	Agricultural and forestry machines—Pedestrian controlled tractors with mounted rotary cultivators, motor hoes, motor hoes with drive wheel(s)—Safety	1997
CEN	EN 710	Safety requirements for foundry moulding and coremaking machinery and plant and associated equipment	1997
CEN	EN 746-1	Industrial thermoprocessing equipment—Part 1: Common safety requirements for industrial thermoprocessing equipment	1997
CEN	EN 746-2	Industrial thermoprocessing equipment—Part 2: Safety requirements for combustion and fuel handling systems	1997

CEN	EN 746-3	Industrial thermoprocessing equipment—Part 3: Safety requirements for the generation and use of atmosphere gases	1997
CEN	EN 774	Garden equipment—Hand held, integrally powered hedge trimmers—Safety	1996
CEN	EN 774/A1	Garden equipment—Hand held, integrally powered hedge trimmers—Safety	1997
CEN	EN 774/A2	Garden equipment—Hand held, integrally powered hedge trimmers—Safety	1997
CEN	EN 775	Manipulating industrial robots—Safety (ISO 10218: 1992, modified)	1992
CEN	EN 786	Garden equipment—Electrically powered walk—behind and lawn edge trimmers—Mechanical safety	1996
CEN	EN 791	Drill rigs—Safety	1995
CEN	EN 811	Safety of machinery—Safety distances to prevent danger zones being reached by the lower limbs	1996
CEN	EN 815	Safety of unshielded tunnel boring machines and rodless shaft boring machines for rock	1996
CEN	EN 818-1	Short link chain for lifting purposes—Safety—Part 1: General conditions of acceptance	1996
CEN	EN 818-2	Short link chain for lifting purposes—Safety—Part 2: Medium tolerance chain for chain slings—Grade 8	1996
CEN	EN 818-4	Short link chain for lifting purposes—Safety—Part 4: Chain slings—Grade 8	1996
CEN	EN 836	Garden equipment—Powered lawnmowers—Safety	1997
CEN	EN 836/A1	Garden equipment—Powered lawnmowers—Safety	1997
CEN	EN 842	Safety of machinery—Visual danger signals—General requirements, design and testing	1996
CEN	EN 859	Safety of woodworking machines—Handfed surface planing machines	1997
CEN	EN 860	Safety of woodworking machines—One side thickness planing machines	1997
CEN	EN 861	Safety of woodworking machines—Surface planing and thicknessing machines	1997

Appendix II: List of Standards • 235

CEN	EN 869	Safety requirements for high pressure metal diecasting units	1997
CEN	EN 894-1	Safety of machinery—Ergonomics requirements for the design of displays and control actuators—Part 1: General principles for human interactions with displays and control actuators	1997
CEN	EN 894-2	Safety of machinery—Ergonomics requirements for the design of displays and contro l actuators—Part 2: Displays	1997
CEN	EN 907	Agricultural and forestry machinery—Sprayers and liquid fertilizers distributors—Safety	1997
CEN	EN 930	Footwear, leather and imitation leather goods manufacturing machines—Roughing, scouring, polishing and trimming machines—Safety requirements	1997
CEN	EN 931	Footwear manufacturing machines—Lasting machines—Safety requirements	1997
CEN	EN 940	Safety of woodworking machines—Combined woodworking machines	1997
CEN	EN 953	Safety of machinery—Guards—General requirements for the design and construction of fixed and movable guards	1997
CEN	EN 954-1	Safety of machinery—Safety—related parts of control systems—Part 1: General principles for design	1996
CEN	EN 981	Safety of machinery—System of auditory and visual danger and information systems	1996
CEN	EN 982	Safety of machinery—Safety requirements for fluid power systems and their com ponents—Hydraulic	1996
CEN	EN 983	Safety of machinery—Safety requirements for fluid power systems and their components —Pneumatics	1996
CEN	EN 996	Pilling equipment—Safety requirements	1995
CEN	EN 1012-1	Compressors and vacuum pumps—Safety requirements—Part 1: compressors	1996
CEN	EN 1012-2	Compressors and vacuum pumps—Safety requirements—Part 2: Vacuum pumps	1996
CEN	EN 1032	Mechanical vibration—Testing of mobile machinery in order to determine the whole—body vibration emission value—General	1996

CEN	EN 1033	Hand—arm vibration—Laboratory measurement of vibration at the grip surface of hand—guided machinery—General	1995
CEN	EN 1037	Safety of machinery—Prevention of unexpected start—up	1995
CEN	EN 1050	Safety of machinery—Principles for risk assessment	1996
CEN	EN 1088	Safety of machinery—Interlocking devices associated with guards—Principles for design and selection	1995
CEN	EN 1093-3	Safety of machinery—Evaluation of the emission of airborne substances—Part 3: Emission rate of a specified pollutant—Bench test method using the real pollutant	1996
CEN	EN 1093-4	Safety of machinery—Evaluation of the emission of airborne substances—Part 4: Capture efficiency of an exhaust system—Tracer method	1996
CEN	EN 1114-1	Rubber and plastic machines—Extruders and extrusion lines—Part 1: Safety requirements for extruders	1996
CEN	EN 1127-1	Explosive atmospheres—Explosion prevention and protection—Part 1: Basic concepts and methodology	1997
CEN	EN 1152	Tractors and machinery for agriculture and forestry—Guards for power take—off (PTO) drive shafts—Wear and strength tests	1994
CEN	EN 1175-2	Safety of industrial trucks—Electrical requirements—Part 2: General requirements of internal combustion engine powered trucks	1998
CEN	EN 1299	Mechanical vibration and shock—Vibration isolation of machines—Information for the application of source isolation	1997
CEN	EN 1398	Dock levellers	1997
CEN	EN 1417	Rubber and plastic machines—Two roll mills—Safety requirements	1996
CEN	EN 1454	Portable, hand—held, internal combustion cutting—off machines—Safety	1997
CEN	EN 1495	Lifting platforms—Mast climbing work platforms	1997
CEN	EN 1525	Safety of industrial trucks—Driverless trucks and their systems	1997

Appendix II: List of Standards

CEN	EN 1526	Safety of industrial trucks—Additional requirements for automated functions on trucks	1997
CEN	EN 1550	Machine—tools safety—Safety requirements for the design and construction of work holding chucks	1997
CEN	EN 1612-1	Rubber and plastic machines—Reaction moulding machines—Part 1: Safety requirements for metering and mixing units	1997
CEN	EN 1672-2	Food processing machinery—Basic concepts—Part 2: Hygiene requirements	1997
CEN	EN 1679-1	Reciprocating internal combustion engines—Safety—Part 1: Compression ignition engines	1998
CEN	EN 1760-1	Safety of machinery—Pressure sensitive protective devices—Part 1: General principles for the design and testing of pressure sensitive mats and pressure sensitive floors	1997
CEN	EN ISO 3450	Earth—moving machinery—Braking systems of rubber—tyred machines—Performance requirements and test procedures (3450:1995)	1996
CEN	EN ISO 3457	Earth—moving machinery—Guards and shields—Definitions and specifications (ISO 3457:1986)	1995
CEN	EN ISO 3743-1	Acoustic—Determination of sound power levels of noise sources—Engineering methods for small, movable sources in reverberant fields—Part 1: Comparison method for hard—walled test rooms (ISO 3743-1:1994)	1995
CEN	EN ISO 3743-2	Acoustic—Determination of sound power levels of noise sources—Engineering methods for small, movable sources in reverberant fields—Part 2: Methods for special reverberation test rooms (ISO 3743-2:1994)	1996
CEN	EN ISO 3744	Acoustic—Determination of sound power levels of noise sources using sound pressure—Engineering method in an essentially free field over a reflecting plane (ISO 3744:1994)	1995
CEN	EN ISO 3746	Acoustic—Determination of sound power levels of noise sources using sound pressure—Survey method using an enveloping measurement surface over a reflecting plane (ISO 3746:1995)	1995

CEN	EN ISO 3767-1	Tractors, machinery for agriculture and forestry, powered lawn and garden equipment—Symbols for operator controls and other displays—Part 1: Common symbols (ISO 3767-1:1991)	1995
CEN	EN ISO 3767-2	Tractors, machinery for agriculture and forestry, powered lawn and garden equipment—Symbols for operator controls and other displays—Part 2: Symbols for agricultural tractors and machinery (ISO 3767-2:1991)	1995
CEN	EN ISO 3767-3	Tractors, machinery for agriculture and forestry, powered lawn and garden equipment—Symbols for operator controls and other displays—Part 3: Symbols for powered lawn and garden equipment (ISO 3767-3:1995)	1996
CEN	EN ISO 3767-4	Tractors, machinery for agriculture and forestry, powered lawn and garden equipment—Symbols for operator controls and other displays—Part 4: Symbols for forestry machinery (ISO 3767-4:1993)	1995
CEN	EN ISO 3767-5	Tractors, machinery for agriculture and forestry, powered lawn and garden equipment—Symbols for operator controls and other displays—Part 5: Symbols for manual portable forestry machinery (ISO 3767-5:1992)	1995
CEN	EN ISO 4871	Acoustics—Declaration and verification of noise emission values of machinery and equipment (ISO 4871:1996)	1996
CEN	EN ISO 6682	Earth—moving machinery—Zones of comfort and reach for controls (ISO 6682:1986 including Amendment 1:1989)	1995
CEN	EN ISO 7235	Acoustics—Measurement procedures for ducted silencers—Insertion loss, flow noise and total pressure loss (ISO 7235:1991)	1995
CEN	EN ISO 7250	Basic human body measurements for technological design (ISO 725:1996)	1997
CEN	EN ISO 8230	Safety requirements for dry—cleaning machines using perchloroethylene	1997
CEN	EN ISO 8662-4	Hand—held portable power tools—Measurement of vibrations at the handle—Part 4: Grinders (ISO 8662-4:1994)	1995

Appendix II: List of Standards

CEN	EN ISO 8662-6	Hand—held portable power tools—Measurement of vibrations at the handle—Part 6: Impact drills (ISO 8662-6:1994)	1995
CEN	EN ISO 8662-7	Hand—held portable power tools—Measurement of vibrations at the handle—Part 7: Wrenches, screwdrivers and nut runners with impact, impulse or ratchet action (ISO 8662-7:1997)	1997
CEN	EN ISO 8662-8	Hand—held portable power tools—Measurement of vibrations at the handle—Part 8: Polishers and rotary, orbital and random orbital sanders (ISO 8662-8:1997)	1997
CEN	EN ISO 8662-9	Hand—held portable power tools—Measurement of vibrations at the handle—Part 9: Rammers (ISO 8662-9:1996)	1996
CEN	EN ISO 8662-12	Hand—held portable power tools—Measurement of vibrations at the handle—Part 12: Saws and files with reciprocating action and saws with oscillating or rotating action (ISO 8662-12:1997)	1997
CEN	EN ISO 8662-13	Hand—held portable power tools—Measurement of vibrations at the handle—Part 13: Die grinders (ISO 8662-13:1997)	1997
CEN	EN ISO 8662-14	Hand—held portable power tools—Measurement of vibrations at the handle—Part 14: Stone—working tools and needle scalers (ISO 8662-14:1996)	1996
CEN	EN ISO 9614-1	Acoustics—Determination of sound power levels of noise sources using sound intensity—Part 1: Measurement at discrete points (ISO 9614-1:1993)	1995
CEN	EN ISO 10472-1	Safety requirements for industrial laundry machinery—Part 1: Common requirements (ISO 10472-1:1997)	1997
CEN	EN ISO 10472-2	Safety requirements for industrial laundry machinery—Part 2: Washing machines and washer—extractors (ISO 10472-2:1997)	1997
CEN	EN ISO 10472-3	Safety requirements for industrial laundry machinery—Part 3: Washing tunnel lines including component machines (ISO 10472-3:1997)	1997
CEN	EN ISO 10472-4	Safety requirements for industrial laundry machinery—Part 4: Air dryers (ISO 10472-4:1997)	1997

CEN	EN ISO 10472-5	Safety requirements for industrial laundry machinery—Part 5: Flatwork ironers, feeders and folders (ISO 10472-5:1997)	1997
CEN	EN ISO 10472-6	Safety requirements for industrial laundry machinery—Part 6: Ironing and fusing presses (ISO 10472-6:1997)	1997
CEN	EN ISO 11102-1	Reciprocating internal combustion engines—Handle starting equipment—Part 1: Safety requirements and tests (ISO 11102-1:1997)	1997
CEN	EN ISO 11102-2	Reciprocating internal combustion engines—Handle starting equipment—Part 2: Method of testing the angle of disengagement (ISO 11102-2:1997)	1997
CEN	EN ISO 11111	Safety requirements for textile machinery (ISO 11111:1995)	1995
CEN	EN ISO 11145	Optic and optical instruments—Lasers and laser—related equipment—Vocabulary and symbols (ISO 11145:1994)	1994
CEN	EN ISO 11200	Acoustics—Noise emitted by machinery and equipment—Guidelines for the use of basic standards for the determination of emission sound pressure level at a work station and at other specified positions (ISO 11200:1995)	1995
CEN	EN ISO 11201	Acoustics—Noise emitted by machinery and equipment—Measurement of emission sound pressure levels at a work station and at other specified positions—Engineering method in an essentially free field over a reflecting plane (ISO 11201:1995)	1995
CEN	EN ISO 11202	Acoustics—Noise emitted by machinery and equipment—Measurement of emission sound pressure levels at a work station and at other specified positions—Survey method in situ (ISO 11202:1995)	1995
CEN	EN ISO 11203	Acoustics—Noise emitted by machinery and equipment—Determination of emission sound pressure levels at a work station and at other specified positions from the sound power level (ISO 11203:1995)	1995
CEN	EN ISO 11204	Acoustics—Noise emitted by machinery and equipment—Measurement of emission sound pressure levels at a work station and at other specified positions—Method requiring environmental corrections (ISO 11204:1995)	1995

CEN	EN ISO 11546-1	Acoustics—Determination of sound insulation performances of enclosures—Part 1: Measurements under laboratory conditions (for declaration purposes) (ISO 11546-1:1995)	1995
CEN	EN ISO 11546-2	Acoustics—Determination of sound insulation performances of enclosures—Part 2: Measurements in situ (for acceptance and verification purposes) (ISO 11546-2:1995)	1995
CEN	EN ISO 11691	Acoustics—Measurement of insertion loss of ducted silencers without flow—Laboratory survey method (ISO 11691:1995)	1995
CEN	EN ISO 11806	Agricultural and forestry machinery—Portable hand—held combustion engine driven brush cutters and grass trimmers—Safety (ISO 11806:1997)	1997
CEN	EN ISO 11957	Acoustics—Determination of sound insulation performance of cabins—Laboratory and in sity measurements (ISO 11957:1996)	1996
CEN	EN ISO 12001	Acoustics—Noise emitted by machinery and equipment—Rules for the drafting and presentation of a noise test code (ISO 21001:1996)	1996
CEN	EN ISO 12626	Safety of machinery—Laser processing machines—Safety requirements (ISO 11553:1996 modified)	1997
CEN	EN ISO 12643	Earth—moving machinery—Rubber—tyred machines—Steering requirements (ISO 5010:1992 modified)	1997
CEN	EN 23741	Acoustics—Determination of sound power levels of noise sources—Precision methods for broadband sources in reverberation rooms (ISO 3741:1988)	1991
CEN	EN 23742	Acoustics—Determination of sound power levels of noise sources—Precision method for discrete—frequency and narrowband sources in reverberation rooms (ISO 3742:1988)	1991
CEN	EN 25136	Acoustics—Determination of sound power radiated into a duct by fans—In-duct method (ISO 5136:1990 and technical corrigendum 1:1993)	1993
CEN	EN 28094	Steel cord conveyor belts—Adhesion strength test of the cover to the core layer (ISO 8094:1984)	1994

CEN	EN 28662-1	Hand—held portable power tools—Measurement of vibrations at the handle—Part 1: General (ISO 8662-1:1988)	1992
CEN	EN 28662-2	Hand—held portable power tools—Measurement of vibrations at the handle—Part 2: Chipping hammers and riveting hammers (ISO 8662-2:1992)	1994
CEN	EN 28662-2/A1	Hand—held portable tools—Measurement of vibrations at the handle—Part 2: Chipping hammers and riveting hammers (ISO 8662-2:1992)	1995
CEN	EN 28662-3	Hand—held portable tools—Measurement of vibrations at the handle—Part 3: Rock drills and rotary hammers (ISO 8662-3:1992)	1994
CEN	EN 28662-3/A1	Hand—held portable tools—Measurement of vibrations at the handle—Part 3: Rock drills and rotary hammers (ISO 8662-3:1992)	1995
CEN	EN 28662-5	Hand—held portable tools—Measurement of vibrations at the handle—Part 5: Pavement breakers and hammers for constuction work (ISO 8662-5:1992)	1994
CEN	EN 28662-5/A1	Hand—held portable power tools—Measurement of vibrations at the handle—Part 5: Pavement breakers and hammers for constuction work (ISO 8662-5:1992)	1995
CEN	EN 30326-1	Mechanical vibration—Laboratory method for evaluating vehicle seat vibration—Part 1: Basic requirements (ISO 10326-1:1992)	1994
CEN	EN 31252	Laser and laser—related equipment—Laser device—Minimum requirements for documentation (ISO 11252:1993)	1994
CEN	EN 31253	Laser and laser—related equipment—Laser device—Mechanical interfaces (ISO 11253:1993)	1994
CLC	EN 60204-1	Safety of machinery—Electrical equipment of machines—Part 1: General requirements	1992

Appendix III Checklist

1. PERSON, PRODUCT, AND ENVIRONMENT

Person

- Age
- Gender
- Experience
- Knowledge
- Habits
- Education
- Familiarity with product
- Physical dimensions and ratios
    - Physical sizes and discrepancies
    - Distances of reach (extremes)
    - Centres of gravity, body parts
    - Complaints
    - Handicaps
- Endurance
    - Condition
    - Health complaints
- Left-handed or right-handed
- Use of medicines
- Pregnancy
- Clothing
    - Tight or loose
    - Footwear heel, slippery, steel toe, bare feet
    - Hairstyle
- Motivation
    - Aim, effort
    - Haste
    - Concentration
    - Mental capacity
- User's manuals
    - Required manuals
    - Possible manuals

- Sequence of manuals
- Posture
- Exercise of power
- Duration of use
- Sight
    - Eye discrepancies, glasses wearer
    - Sight of product
    - Sight of environment

Product

- Mechanical dangers
    - Dimensions
    - Static openings
    - Sharp corners, points
    - Sharp edge
    - Air permeability
    - Surface
- Potential energy
    - Pushing away
    - Falling over
    - Lifting of persons
    - Strength or stiffness insufficient
    - Elastic element
    - Gas or liquid under high pressure
    - High operating position for person
- Kinetic energy
    - Parts collide
    - Part move along each other
    - Rotating parts
    - Acceleration, retardation
    - Falling objects
    - Transport of persons
    - Objects shooting out
    - Noise
    - Shock wave
    - Vibration
    - Material fatigue

- Electricity
  - Voltage
  - Current
  - Defective insulation
  - Static electricity
  - Cause or react to magnetic field
  - Short circuit
- Temperature
  - Open flame
  - Cold or hot surface
  - Cold or hot liquid
  - Cold or hot gas
- Radiation
  - Laser
  - Infrared
  - Visible high-intensity light
  - Ultraviolet radiation
  - Ionising radiation
  - Microwaves
  - Magnetism
- Fire and explosion
  - Concentration of oxygen
  - Fuel
  - High temperatures
  - Oxidation
  - Explosive material
- Chemical
  - Solid matter or liquid
  - Steam
  - Corrosive product
  - Corrosive gas or vapor
  - Immersion in toxic liquid
  - Bury under grains
- Power supply
  - Interruption or failure
  - Defective
  - Will not start
  - Will not stop

Environment

- Physical environment
    - Light
    - Noise
    - Temperature
    - Moisture
    - Air pollution
    - Airflow
    - Dust production
    - Vibrations
    - Gases, liquids, or vapors present
    - Weather conditions
    - Bacteria, algae
    - Vegetation
    - Height of operation (fear of heights)
    - Oxygen content in air
    - Place of operation
    - Setup of room of use
    - Underground
    - Material to be processed
    - Other products
    - Wrong or unclear instructions
    - Presence of animals
- Social environment
    - Cultural aspects
    - Economic values
    - Social structures
    - Patterns of expectations
    - Persons present
    - Persons absent

2. PROCESS TREE

In order to aid the systematic identification of latent risks, the risk evaluation is divided into categories which match the separate phases in the life cycle of the product. It is recommended that the process tree first be developed completely before putting the risks on paper. Below is a process tree showing, for the purposes of illustration, the life cycle of an electric stapler.

Process tree electric stapler

1. Origin
- Design
  - Draw design
  - Build model
  - Provide production documentation
- Prepare for production
  - Purchased material
  - Purchase machines
  - Set up machines
  - Make tools and moulds
- Production
  - Die-case parts
  - Make printing plate
  - Lay out parts
  - Assemble parts
- Packaging
  - Package in film
  - Lay in unit box
  - Stack in large box
  - Place on pallet

2. Distribution
- Transport
  - Lift with forklift truck
  - Place on truck
  - Unload
- Storage
  - Place pallet in warehouse
  - Open large box
  - Place product with packaging in shop window

- Sale
    - Hand over to customer
    - Take home with him or her
    - Store unit box in closet

3. Use
- Make ready for use
    - Open unit box
    - Remove product and examine
    - Remove magazine
    - Insert staples
    - Close magazine
    - Connect plug to socket
- Use
    - Adjust power
    - Position stapler against part to be secured
    - Align staple
    - Fire staple
- Cleaning
    - Wipe with cloth
    - Blow away dust
- Repair
    - Unscrew screws
    - Dismantle housing
    - Change parts

4. Inappropriate use
- Tampering with safety features
    - Pushing the magazine up by hand
    - Firing staple into the air

5. Disposal
- Dismantle
    - Unscrew screws
    - Dismantle product
    - Separate parts according to material

- Process waste
  - Deposit material in designated waste bin
  - Possibly return parts to the manufacturer

3. OTHER RESOURCES

- Fundamental requirements from the relevant Directive
- Standards, possible reference by Directive
- Relevant industrial regulations and standards
- Drawings of the machine or other means of reproducing the expected features
- Computer simulations
- Details of all desired and possible operations, including startup, use, emergencies, repair, fault location, cleaning, and use of personal protective clothing
- Details of potentially hazardous materials which are used in manufacturing the product (or which are processed by it in the case of a machine)
- Details about transport, installation, testing, production, dismantling, and disposal
- Accident data; also, for related products for figures relating to accident registration, help can be obtained from, among other:
  - SCV (Stichting Consument en Veiligheid [Consumers and Safety Institute])
  - CBS (Centraal Bureau voor de Statistiek [Central Bureau of Statistics])
  - NIA (Nederlands Instituut voor Arbeidsomstandigheden [Netherlands Institute for Working Conditions])
  - SWOV (Stichting Wetenschappelijk Onderzoek Verkeersveiligheid [Institute for Scientific Research into Traffic Safety])
  - NBS (Nederlandse Brandwondenstichting [Netherlands Burns Institute])
  - IGB (Inspectie Gezondheidsbescherming [Health and Safety Inspectorate])
  - DTI (Department of Trade and Industry, UK)
  - CPSC (Consumer Product Safety Commission, USA)

- Reliability information on components, systems, and human intervention
- Information on the environment in which the product will be used
- Indication of the level of training and experience of the user
- Examination of use
- Information on similar products
- Other sources of information:
  - Consumers' association
  - ANWB (Algemene Nederlandse Wielrijdersbond [General Netherlands Cyclists' Association])
  - ENFB (Eerste Nederlandse Fietsers Bond [First Dutch Cyclists' Association])
  - Ministry of Health, Welfare and Sport
  - Labour Inspectorate (Social Affairs and Employment)

Appendix

IV

Addresses with Regard to
the Machinery Directive

Belgium

AIB-Vinçotte Inter
Avenue André-Drouart 27-29
B-160 BRUXELLES

Denmark

Demko
Lyskaer 8
Postboks 514
DK-2730 HERLEV

Dansk Teknologisk Institut
Industriel Teknologi
Teknologiparken
DK-8000 ÅRHUS C

Germany

Fachausschuß Chemie
Prüf- und Zertifizierungsstelle im BG-Prüfzert
Kurfürsten-Anlage 62
D-69115 HEIDELBERG

Fachausschuß Tiefbau
Prüf- und Zertifizierungsstelle im BG-Prüfzert
Am Knie 6
D-81241 MÜNCHEN

Zertifizierungsstelle für Qualitätssicherungssysteme und Produkte des Staatlichen Materialprüfungsamtes MPA-NRW
Marsbruchstraße 186
D-44287 DORTMUND

Technische Prüfungsstelle Dienstbier & Pix
Hundert Beete 13
D-91334 HEMHOFEN

TÜV Südwestdeutschland e.v.
TÜV Cert-Zertifizierungsstelle
Dudenstraße 28
D-68167 MANNHEIM

RW TÜV E.V./RW TÜV Anlagentec GmbH
TÜV Cert-Zertifizierungsstelle
Dudenstraße 28
D-68167 MANNHEIM

TüV Nord E.V.
TüV Cert-Zertifizierungsstelle
Große Bahnstraße 31
D-22525 HAMBURG

TüV Bayern/Sachsen E.V.
TüV Cert-Zertifizierungsstelle
Westendstraße 199
D-80686 MÜNCHEN

SLG Prüf- und Zertifizierungs-GmbH
Markt 5
D-09111 CHEMNITZ

Fachausschuß "Verkehr," Prüf- und Zertifizierungsstelle im BG-Prüfzert
Max Brauer Allee 41
D-22765 HAMBURG

Deutsche Prüfstelle für Land- und Forsttechnik (DPLF)
Weissensteinstraße 70/72
D-34114 KASSEL

Fachausschuß Eisen und Metall II und Hebezeug II
Prüf- und Zertifizierungsstelle im BG-Prufzert

Kreuzstraße 45
D-40210 DÜSSELDORF

DMT-Zertifizierungsstelle der DMT-Gesellschaft für Forschung und Prüfung MBH
Franz Fischerweg 61
D-45307 ESSEN

Berufsgenossenschaftliches Institut für Arbeitssicherheit BIA
Alte Heerstraße 111
D-53757 SANKT AUGUSTIN

Fachausschuß Fleischwirtschaft
Prüf- und Zertifizierungsstelle im BG-Prüfzert
Lortzingstraße 2
D-55127 MAINZ

Fachausschuß Holz
Prüf- und Zertifizierungsstelle im BG-Prüfzert
Vollmöllerstraße 11
D-70563 STUTTGART

VDE-Verband Deutscher Elektrotechniker E.V.
VDE-Prüf- und Zertifizierungsinstitut
Merianstraße 28
D-63069 OFFENBACH

Dekra A.G.
Prüf- und Zertifizierungsstelle
Schulze Delitzschstraße 49
D-70565 STUTTGART

TüV Hannover/Sachsen-Anhalt e.v.
TüV Cert-Zertifizierungsstelle
Am TüV 1
D-30519 HANNOVER

Landesgewerbeanstalt Bayern
Prüfstelle fur Geratesicherheit LGA

Tillystraße 2
D-90431 NÜRNBERG

TüV Rheinland Product Safety GMBH
Zertifizierungs- und Prüfstelle fur Gerätesicherheit
Am Grauen Stein
D-51105 KÖLN

TüV Product Service GMBH
Ridlerstraße 31
D-80339 MÜNCHEN

TÜV Saarland e.v.
TÜV Cert-Zertifizierungsstelle
Saarbruckerstraße 8
D-66280 SULZBACH

TÜV Berlin-Brandenburg e.v.
TÜV Cert-Zertifizierungsstelle
Magirusstraße 5
D-12103 BERLIN

Finland

Agricultural Research Centre of Finland
Institute of Agricultural Engineering
MTT/Vakola, Inspection Office
Vakolantie 55
FIN-03400 VIHTI

Teknillinen Tarkastuskeskus TKK
P.O. Box 204 (Lönnrotinkatu 37)
FIN-00181 HELSINKI

Electrial Inspectorate
P.O. Box 21
FIN-00211 HELSINKI

VTT Valmistustekniikka/Turvallisuustekniikka
(VTT Manufacturing Technology/Safety Engineering)
P.O. Box 17011
FIN-33101 TAMPERE

France

Centre Technique des Industries Mécaniques (CETIM)
52 Rue Felix-Louat
B.P. 67
F-60304 SENLIS

Apave Lyonnaise
B.P. 3
F-69611 TASSIN CEDEX

Association Interprofessionelle de France (AINF)
Zone Industrielle
B.P. 259
F-59472 SECLIN CEDEX

Apave Alsacienne
2 Rue Thiers
B.P. 1347
F-68056 MULHOUSE CEDEX

Centre National du Machinisme Agricole, du Genie rural, des Eaux et Fôrets
(Cemagref)
Parc de Tourvoie
B.P. 121
F-92185 ANTONY CEDEX

Apave du Sud-Ouest
Zone Industrielle
B.P. 3
F-33370 TRESSES CEDEX

Apave Normande
2 Rue des Mouettes
F-76130 MONT-SAINT-AIGNAN

Apave Nord-Picardie
B.P. 247
F-59019 LILLE CEDEX

Contrôle et Prévention
34 Rue Rennequin
F-75850 PARIS CEDEX 17

Centre Technique du Bois et de L'Ameublement (CTBA)
10 Avenue de Saint-Mande
F-70512 PARIS

AIF Services SA (AIF)
Zone Industrielle de Magre
B.P. 308
F-87008 LIMOGES CEDEX

Assocation pour le Developpement de l'Institut de la Viande
B.P.-Saint-Jean
F-63015 CLERMONT-FERRAND CEDEX

Apave de l'Ouest
B.P. 289
F-44803 SAINT-HERBLAIN CEDEX

Apave Parisienne
13-17 Rue Salneuve
F-75854 PARIS CEDEX 17

Institut National de l'Environnement Industriel et des Risques (Ineris)
B.P. 2
F-60550 VERNEUIL-EN-HALATTE

Union Technique de l'Automobile, du Motocycle et du Cycle
Autodrome de Linas
F-91310 MONTLHÉRY

Laboratoire National d'Essais (LNE)
1 Rue Gaston-Boissier
F-75015 PARIS

Institut National de Recherche et de Sécurité INRS
B.P. 27
F-54501 VANDCEUVRE CEDEX

Ireland

National Standards Authority of Ireland
NSAI
Glasnevin
IRL-DUBLIN 9

Italy

Istedil Spa-Instituto Sperimentale per L'Edilizia
Via Tiburtina KM 18.300
I-00012 SETTEVILLE DIE GUIDONIA (RM)

Instituto di Certificazione Industriale per la Meccanica ICIM
Via Giardino 4
I-20123 MILANO

Instituto Italiano del Marchio di Qualita IMQ
Via Quintiliano 43
I-20138 MILANO

Instituto di Certificazione Europea
Prodotti Industriali SRL ICEPI
Via Emilia Parmense I IA
I-29010 Pontenure (PC)

Cermet
Via Aldo Moro 22
I-40068 S. LAZZARO DI SAVENA (BO)

Modulo Uno
Via Cuorgne 21
I-10100 TORINO

Industrial Engineering Consultant IEC
Via Botticelli 51
I-10154 TORINO

Instituto Certificazione Europea SRL
ICE BO
Via Bentini 9
I-40013 CASTEL MAGGIORE (BO)

Instituto Giordano Spa
Via Rossini 2
I-47041 BELLARIA - IGEA MARINA (RN)

Instituto Recerche, Prove ed Analisi CPM
Via Artigiani 63
I-25040 BIENNO (BS)

Organismo Certificazione Europea OCE
Via Ancona 21
I-00198 ROMA

Agenzia Nazionale Certificazione Componenti in Pressione
ANCCP
Via Bronzino 3
I-20123 MILANO

Ing. Cesare Petrosillo (Petrosillo Engineering Group SRL)
Via Madre Grazie 12
1-74100 TARANTO

Instituto di Ricerche e Collaudi M. Masini SRL
Via Moscova 11
I-20017 RHO (MILANO)

The Netherlands

Stichting TNO Certification TNO
Laan van Westenenk 501
Postbus 541
NL-7300 AM APELDOORN

Stichting Keuringsbureau Hout SKH
Huizermaatweg 29
NL-1273 NAHUIZEN

Stichting Aboma + Keboma
Pascalstraat 9
Postbus 141
NL-6710 BC EDE

Stichting Nederlands Instituut voor Lifttechniek
Liftinstituut
Buikslotermeerplein 381
Postbus 36027
NL-1020 MA AMSTERDAM

KEMA N.V.
Utrechtseweg 310
Postbus 9035
NL-6800 ET ARNHEM

Norway

Det Norske Veritas Industry AS DNV
P.O. Box 300
N-1322 HOVIK

Norwegian Board for Testing and Approval of Electric Equipment
NEMKO
P.O. Box 73 Blindern
N-0314 OSLO

Austria

Technischer Überwachungsverein Österreich
TÜV A
Krügerstraße 16
A-1015 WIEN

Portugal

Centro de Apoio à Indústria Metalomecanica
CATIM
Rua dos Platanos 197
P-4100 PORTO

Instittuto de Soldadura e Qualidade (ISQ)
EN 249-KM 3, Cabanas-Leiao (Tagus-park)
Apartado 119
P-2781 OEIRAS CODEX

Spain

Entidad Colaboradora de la Administración SA
ECA
Travessera de Dalt 130- 134
E-08024 BARCELONA

Eurocontrol SA
Zurbano 48
E-28010 MADRID

Centor de Inspección y Asistencia Técnica, SA CIAT
Piquer 7
E-28033 MADRID

Inspección y Garantia de Calidad SA IGC
CTRO Empresarial Atica 7
Avda de Europa 26 - Ed.5.PL.1 A
E-28224 POZUELO DE ALARCON

ICICT SA
Buenavista 30
E-08012 BARCELONA

Lloyd's Register of Shipping
Princesa 29
E-28008 MADRID

Norcontrol SA
Francisco Gervas 14-1B
E-28020 MADRID

Novotec Consultores, SA
Colombia 62
E-28016 MADRID

Tecnos, Garantia de Calidad SA
Mesena 39
E-28033 MADRID

Asociación Española de Normalización y Certificación Aenor
Fernández de la Hoz 52
E-28010 MADRID

Estación Mecánica Agricola
EMA
Carretera de Toledo, KM 6.800
E-28916 LEGANÉS

Laboratorios del Area de Verificacion de Maquineria
CTRO Nacional de Verificacion de Maquineria de Vizcaya
Camino de la Dinamita S/N
E-48903 BARAKALDO

Laboratorio Oficial José Maria De Madariaga
LOM
Alenza 1-2
E-28003 MADRID

ACI, SA
Duque de Sesto 34
E-28009 MADRID

Asistencia Técnica Industrial SAE
ATISAE
San Telmo 28
E-28016 MADRID

Bureau Veritas Español SA
Dr. Fleming 31
E-28036 MADRID

Cualicontrol SA
Juan Bautista de Toledo 31
E-28002 MADRID

United Kingdom

AMTRI Veritas Limited
Hulley Road
UK-SK10 2NE MACCLESFIELD CHESHIRE

ERA Technology Ltd
Cleeve Road
UK-HT22 7SA LEATHERHEAD SURREY

Lloyd's Register of Shipping
Lloyd's Register House
29 Wellesley Road
UK-CRO 2AJ CROYDON

Plant Safety Ltd
825A Wilmslow Road, Didsbury
UK-M20 ORE MANCHESTER

United Kingdom Atomic Energy Authority
Machinery Certification Service

Thomson House, Risley
UK-WA3 OAT WARRINGTON CHESHIRE

SGS United Kingdom Ltd
SGS House, Johns Lane, Trividale
UK-B69 3HX WARLEY, WEST MIDLANDS

British Standards Institution Testing
Mayland Avenue
UK-HP2 4SQ HEMEL HEMPSTEAD HERTFORDSHIRE

Sweden

AB Svensk Anläggningsprovning (SA)
Box 49306
S-100 29 STOCKHOLM

Statens Maskinprovningar
SMP
Box 7035
S-750 07 UPPSALA

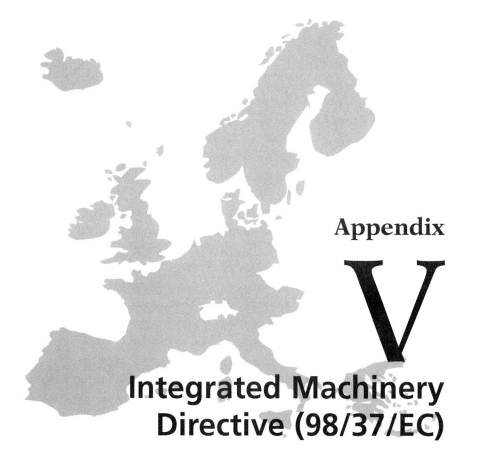

Appendix V

Integrated Machinery Directive (98/37/EC)

This appendix contains the text of the Integrated Machinery Directive 98/37/EEC.

DIRECTIVE 98/37/EC OF THE EUROPEAN PARLIAMENT AND OF THE COUNCIL

of 22 June 1998 on the approximation of the laws of the Member States relating to machinery

THE EUROPEAN PARLIAMENT AND THE COUNCIL OF THE EUROPEAN UNION,

Having regard to the Treaty establishing the European Community, and in particular Article 100a thereof,

Having regard to the proposal from the Commission,

Having regard to the opinion of the Economic and Social Committee,[1]

Acting in accordance with the procedure laid down in Article 189b of the Treaty,[2]

(1) Whereas Council Directive 89/392/EEC of 14 June 1989 on the approximation of the laws of the Member States relating to machinery[3] has been frequently and substantially amended; whereas for reasons of clarity and rationality the said Directive should be consolidated;

(2) Whereas the internal market consists of an area without internal frontiers within which the free movement of goods, persons, services and capital is guaranteed;

---

[1] OJ C 133, 28.4.1997, p. 6.
[2] Opinion of the European Parliament of 17 September 1997 (OJ C 304, 6.10.1997, p. 79), Council common position of 24 March 1998 (OJ C 161, 27.5.1998, p. 54) and Decision of the European Parliament of 30 April 1998 (OJ C 152, 18.5.1998). Council Decision of 25 May 1998.
[3] OJ L 183, 29.6.1989, p. 9. Directive as last amended by Directive 93/68/EEC (OJ L 220, 30.8.1993, p. 1).

(3) Whereas the machinery sector is an important part of the engineering industry and is one of the industrial mainstays of the Community economy;

(4) Whereas the social cost of the large number of accidents caused directly by the use of machinery can be reduced by inherently safe design and construction of machinery and by proper installations and maintenance;

(5) Whereas Member States are responsible for ensuring the health and safety on their territory of persons and, where appropriate, of domestic animals and goods and, in particular, of workers, notably in relation to the risks arising out of the use of machinery;

(6) Whereas, in the Member States, the legislative systems regarding accident prevention are very different; whereas the relevant compulsory provisions, frequently supplemented by de facto mandatory technical specifications and/or voluntary standards, do not necessarily lead to different levels of health and safety, but nevertheless, owing to their disparities, constitute barriers to trade within the Community; whereas, furthermore, conformity certification and national certification systems for machinery differ considerably;

(7) Whereas existing national health and safety provisions providing protection against the risks caused by machinery must be approximated to ensure free movement on the market of machinery without lowering existing justified levels of protection in the Member States; whereas the provisions of this Directive concerning the design and construction of machinery, essential for a safer working environment, shall be accompanied by specific provisions concerning the prevention of certain risks to which workers can be exposed at work, as well as by provisions based on the organisation of safety of workers in the working environment;

(8) Whereas Community law, in its present form, provides-by way of derogation from one of the fundamental rules of the Community, namely the free movement of goods-that obstacles to movement within the Community resulting from disparities in national legislation relating to the marketing of products must be accepted in so far as the provisions concerned can be recognized as being necessary to satisfy imperative requirements;

(9) Whereas paragraphs 65 and 68 of the White Paper on the completion of the internal market, approved by the European Council in June 1985, provide for a new approach to legislative harmonisation; whereas,

therefore, the harmonisation of laws in this case must be limited to those requirments necessary to satisfy the imperative and essential health and safety requirements relating to machinery; whereas these requirements must replace the relevant national provisions because they are essential;

(10) Whereas the maintenance or improvement of the level of safety attained by the Member States constitutes one of the essential aims of this Directive and of the principle of safety as defined by the essential requirements;

(11) Whereas the field of application of this Directive must be based on a general definition of the term "machinery" so as to allow the technical development of products; whereas the development of complex installations and the risks they involve are of an equivalent nature and their express inclusion in the Directive is therefore justified;

(12) Whereas it is also necessary to deal with safety components which are placed on the market separately and the safety function of which is declared by the manufacturer or his authorised representative established in the Community;

(13) Whereas, for trade fairs, exhibitions, etc., it must be possible to exhibit machinery which does not conform to this Directive; whereas, however, interested parties should be properly informed that the machinery does not conform and cannot be purchased in that condition;

(14) Whereas the essential health and safety requirements must be observed in order to ensure that machinery is safe; whereas these requirements must be applied with discernment to take account of the state of the art at the time of construction and of technical and economic requirements;

(15) Whereas the putting into service of machinery within the meaning of this Directive can relate only to the use of the machinery itself as intended by the manufacturer; whereas this does not preclude the laying-down of conditions of use external to the machinery, provided that it is not thereby modified in a way not specified in this Directive;

(16) Whereas it is necessary not only to ensure the free movement and putting into service of machinery bearing the "CE" marking and having an EC conformity certificate but also to ensure free movement of machinery not bearing the "CE" marking where it is to be incorporated into other machinery or assembled with other machinery to form a complex installation;

(17) Whereas, therefore, this Directive defines only the essential health and safety requirements of general application, supplemented by a number of more specific requirements for certain categories of machinery; whereas, in order to help manufacturers to prove conformity to these essential requirements and in order to allow inspection for conformity to the essential requirements, it is desirable to have standards harmonised at European level for the prevention of risks arising out of the design and construction of machinery; whereas these standards harmonised at European level are drawn up by private-law bodies and must retain their non-binding status; whereas for this purpose the European Committee for Standardisation (CEN) and the European Committee for Electrotechnical Standardisation (Cenelec) are the bodies recognized as competent to adopt harmonised standards in accordance with the general guidelines for cooperation between the Commission and these two bodies signed on 13 November 1984; whereas, within the meaning of this Directive, a harmonised standard is a technical specification (European standard or harmonisation document) adopted by either or both of these bodies, on the basis of a remit from the Commission in accordance with the provisions of Directive 83/189/EEC[4] and on the basis of general guidelines referred to above;

(18) Whereas it was found necessary to improve the legislative framework in order to ensure an effective and appropriate contribution by employers and employees to the standardisation process;

(19) Whereas the Member States' responsibility for safety, health and the other aspects covered by the essential requirements on their territory must be recognized in a safeguard clause providing for adequate Community protection procedures;

(20) Whereas, as is currently the practice in Member States, manufacturers should retain the responsibility for certifying the conformity of their machinery to the relevant essential requirements; whereas conformity to harmonized standards creates a presumption of conformity to

---

[4] Council Directive 83/189/EEC of 28 March 1983 laying down a procedure for the provision of information in the field of technical standards and regulations (OJ L 109, 26.4. 1983, p. 8). Directive as last amended by Commission Decision 96/139/EC (OJ L 32, 10.2. 1996, p. 31).

the relevant essential requirements; whereas it is left to the sole discretion of the manufacturer, where he feels the need, to have his products examined and certified by a third party;

(21) Whereas, for certain types of machinery having a higher risk factor, a stricter certification procedure is desirable; whereas the EC type-examination procedure adopted may result in an EC declaration being given by the manufacturer without any stricter requirement such as a guarantee of quality, EC verification or EC supervision;

(22) Whereas it is essential that, before issuing an EC declaration of conformity, the manufacturer or his authorised representative established in the Community should provide a technical construction file; whereas it is not, however, essential that all documentation be permanently available in a material manner, but it must be made available on demand; whereas it need not include detailed plans of the sub-assemblies used in manufacturing the machines, unless knowledge of these is indispensable in order to ascertain conformity with essential safety requirements;

(23) Whereas, in its communication of 15 June 1989 on a global approach to certification and testing,[5] the Commission proposed that common rules be drawn up concerning a "CE" conformity marking with a single design; whereas, in its resolution of 21 December 1989 on a global approach to conformity assessment,[6] the Council approved as a guiding principle the adoption of a consistent approach such as this with regard to the use of the "CE" marking; whereas the two basic elements of the new approach which must be applied are therefore the essential requirements and the conformity assessment procedures;

(24) Whereas the addressees of any decision taken under this Directive must be informed of the reasons for such a decision and the legal remedies open to them;

(25) Whereas this Directive must not affect the obligations of the Member States concerning the deadlines for transposition and application of the Directives set out in Annex VIII, part B,
HAVE ADOPTED THIS DIRECTIVE:

---

[5] OJ C 231, 8.9. 1989, p. 3, and OJ C 267, 19.10. 1989, p. 3.
[6] OJ C 10, 16.1. 1990, p. 1.

CHAPTER I
SCOPE, PLACING ON THE MARKET AND FREEDOM OF MOVEMENT

ARTICLE 1

1. This Directive applies to machinery and lays down the essential health and safety requirements therefor, as defined in Annex I.

It shall also apply to safety components placed on the market separately.

2. For the purposes of this Directive:

(a) "machinery" means:

- an assembly of linked parts or components, at least one of which moves, with the appropriate actuators, control and power circuits, etc., joined together for a specific application, in particular for the processing, treatment, moving or packaging of a material,
- an assembly of machines which, in order to achieve the same end, are arranged and controlled so that they function as an integral whole,
- interchangeable equipment modifying the function of a machine, which is placed on the market for the purpose of being assembled with a machine or a series of different machines or with a tractor by the operator himself in so far as this equipment is not a spare part or a tool;

(b) "safety components" means a component, provided that it is not interchangeable equipment, which the manufacturer or his authorised representative established in the Community places on the market to fulfil a safety function when in use and the failure or malfunctioning of which endangers the safety or health of exposed persons.

3. The following are excluded from the scope of this Directive:

- machinery whose only power source is directly applied manual effort, unless it is a machine used for lifting or lowering loads,
- machinery for medical use used in direct contact with patients,

- special equipment for use in fairgrounds and/or amusement parks,
- steam boilers, tanks and pressure vessels,
- machinery specially designed or put into service for nuclear purposes which, in the event of failure, may result in an emission of radioactivity,
- radioactive sources forming part of a machine,
- firearms,
- storage tanks and pipelines for petrol, diesel fuel, inflammable liquids and dangerous substances,
- means of transport, i.e. vehicles and their trailers intended solely for transporting passengers by air or on road, rail or water networks, as well as means of transport in so far as such means are designed for transporting goods by air, on public road or rail networks or on water. Vehicles used in the mineral extraction industry shall not be excluded,
- seagoing vessels and mobile offshore units together with equipment on board such vessels or units,
- cableways, including funicular railways, for the public or private transportation of persons,
- agricultural and forestry tractors, as defined in Article 1(1) of Directive 74/150/EEC,[7]
- machines specially designed and constructed for military or police purposes,
- lifts which permanently serve specific levels of buildings and constructions, having a car moving between guides which are rigid and inclined at an angle of more than 15 degrees to the horizontal and designed for the transport of:
  (i) persons;
  (ii) persons and goods;
  (iii) goods alone if the car is accessible, that is to say, a person may enter it without difficulty, and fitted with controls situated inside the car or within reach of a person inside,

---

[7] Council Directive 74/150/EEC of 4 March 1974 on the approximation of the laws of the Member States relating to the type-approval of wheeled agricultural or forestry tractors (OJ L 84, 28.3. 1974, p. 10). Directive as last amended by Decision 95/1/EC, Euratom, ECSC (OJ L 1.1. 1995, p. 1).

- means of transport of persons using rack and pinion rail mounted vehicles,
- mine winding gear,
- theatre elevators,
- construction site hoists intended for lifting persons or persons and goods.

4. Where, for machinery or safety components, the risks referred to in this Directive are wholly or partly covered by specific Community Directives, this Directive shall not apply, or shall cease to apply, in the case of such machinery or safety components and of such risks on the implementation of these specific Directives.

5. Where, for machinery, the risks are mainly of electrical origin, such machinery shall be covered exclusively by Directive 73/23/EEC.[8]

ARTICLE 2

1. Member States shall take all appropriate measures to ensure that machinery or safety components covered by this Directive may be placed on the market and put into service only if they do not endanger the health or safety of persons and, where appropriate, domestic animals or property, when properly installed and maintained and used for their intended purpose.

2. This Directive shall not affect Member States' entitlement to lay down, in due observance of the Treaty, such requirements as they may deem necessary to ensure that persons and in particular workers are protected when using the machinery or safety components in question, provided that this does not mean that the machinery or safety components are modified in a way not specified in the Directive.

3. At trade fairs, exhibitions, demonstrations, etc., Member States shall not prevent the showing of machinery or safety components which do not conform to the provisions of this Directive, provided that

---

[8] Council Directive 73/23/EEC of 19 February 1973 on the harmonization of the laws of Member States relating to electrical equipment designed for use within certain voltage limits (OJ L 77, 26.3. 1973, p. 29). Directive as last amended by Directive 93/68/EEC (OJ L 220, 30.8. 1993, p. 1).

a visible sign clearly indicates that such machinery or safety components do not conform and that they are not for sale until they have been brought into conformity by the manufacturer or his authorised representative established in the Community. During demonstrations, adequate safety measures shall be taken to ensure the protection of persons.

ARTICLE 3

Machinery and safety components covered by this Directive shall satisfy the essential health and safety requirements set out in Annex I.

ARTICLE 4

1. Member States shall not prohibit, restrict or impede the placing on the market and putting into service in their territory of machinery and safety components which comply with this Directive.

2. Member States shall not prohibit, restrict or impede the placing on the market of machinery where the manufacturer or his authorised representative established in the Community declares in accordance with point B of Annex II that it is intended to be incorporated into machinery or assembled with other machinery to constitute machinery covered by this Directive, except where it can function independently.

"Interchangeable equipment," as referred to in the third indent of Article 1(2)(a), must in all cases bear the CE marking and be accompanied by the EC declaration of conformity referred to in Annex II, point A.

3. Member States may not prohibit, restrict or impede the placing on the market of safety components as defined in Article 1(2) where they are accompanied by an EC declaration of conformity by the manufacturer or his authorised representative established in the Community as referred to in Annex II, point C.

ARTICLE 5

1. Member States shall regard the following as conforming to all the provisions of this Directive, including the procedures for checking the conformity provided for in Chapter II:

- machinery bearing the CE marking and accompanied by the EC declaration of conformity referred to in Annex II, point A,
- safety components accompanied by the EC declaration of conformity referred to in Annex II, point C.

In the absence of harmonized standards, Member States shall take any steps they deem necessary to bring to the attention of the parties concerned the existing national technical standards and specifications which are regarded as important or relevant to the proper implementation of the essential safety and health requirements in Annex I.

2. Where a national standard transposing a harmonized standard, the reference for which has been published in the Official Journal of the European Communities, covers one or more of the essential safety requirements, machinery or safety components constructed in accordance with this standard shall be presumed to comply with the relevant essential requirements.

Member States shall publish the references of national standards transposing harmonized standards.

3. Member States shall ensure that appropriate measures are taken to enable the social partners to have an influence at national level on the process of preparing and monitoring the harmonized standards.

ARTICLE 6

1. Where a Member State or the Commission considers that the harmonized standards referred to in Article 5(2) do not entirely satisfy the essential requirements referred to in Article 3, the Commission or the Member State concerned shall bring the matter before the committee set up under Directive 83/189/EEC, giving the reasons therefor. The committee shall deliver an opinion without delay.

Upon receipt of the committee's opinion, the Commission shall inform the Member States whether or not it is necessary to withdraw those standards from the published information referred to in Article 5(2).

2. A standing committee shall be set up, consisting of representatives appointed by the Member States and chaired by a representative of the Commission.

The standing committee shall draw up its own rules of procedure.

Any matter relating to the implementation and practical application of this Directive may be brought before the standing committee, in accordance with the following procedure:

The representative of the Commission shall submit to the committee a draft of the measures to be taken. The committee shall deliver its opinion on the draft, within a time limit which the chairman may lay down according to the urgency of the matter, if necessary by taking a vote.

The opinion shall be recorded in the minutes; in addition, each Member State shall have the right to ask to have its position recorded in the minutes.

The Commission shall take the utmost account of the opinion delivered by the committee. It shall inform the committee of the manner in which its opinion has been taken into account.

ARTICLE 7

1. Where a Member State ascertains that:

- machinery bearing the CE marking, or
- safety components accompanied by the EC declaration of conformity, used in accordance with their intended purpose are liable to endanger the safety of persons, and, where appropriate, domestic animals or property, it shall take all appropriate measures to withdraw such machinery or safety components from the market, to prohibit the placing on the market, putting into service or use thereof, or to restrict free movement thereof.

Member States shall immediately inform the Commission of any such measure, indicating the reason for its decision and, in particular, whether non-conformity is due to:

(a) failure to satisfy the essential requirements referred to in Article 3;
(b) incorrect application of the standards referred to in Article 5(2);
(c) shortcomings in the standards themselves referred to in Article 5(2).

2. The Commission shall enter into consultation with the parties concerned without delay. Where the Commission considers, after this consultation, that the measure is justified, it shall immediately so inform the

Member State which took the initiative and the other Member States. Where the Commission considers, after this consultation, that the action is unjustified, it shall immediately so inform the Member State which took the initiative and the manufacturer or his authorised representative established within the Community. Where the decision referred to in paragraph 1 is based on a shortcoming in the standards, and where the Member State at the origin of the decision maintains its position, the Commission shall immediately inform the committee in order to initiate the procedures referred to in Article 6(1).

3. Where:

- machinery which does not comply bears the CE marking,
- a safety component which does not comply is accompanied by an EC declaration of conformity, the competent Member State shall take appropriate action against whom so ever has affixed the marking or drawn up the declaration and shall so inform the Commission and other Member States.

4. The Commission shall ensure that Member States are kept informed of the progress and outcome of this procedure.

CHAPTER II
CONFORMITY ASSESSMENT PROCEDURES

ARTICLE 8

1. The manufacturer or his authorised representative established in the Community must, in order to certify that machinery and safety components are in conformity with this Directive, draw up for all machinery or safety components manufactured an EC declaration of conformity based on the model given in Annex II, point A or C as appropriate.

In addition, for machinery alone, the manufacturer or his authorised representatives established in the Community must affix to the machine the CE marking.

2. Before placing on the market, the manufacturer, or his authorised representative established in the Community, shall:

(a) if the machinery is not referred to in Annex IV, draw up the file provided for in Annex V;

(b) if the machinery is referred to in Annex IV and its manufacturer does not comply, or only partly complies, with the standards referred to in Article 5(2) or if there are no such standards, submit an example of the machinery for the EC type-examination referred to in Annex VI;

(c) if the machinery is referred to in Annex IV and is manufactured in accordance with the standards referred to in Article 5(2):

- either draw up the file referred to in Annex VI and forward it to a notified body, which will acknowledge receipt of the file as soon as possible and keep it,
- submit the file referred to in Annex VI to the notified body, which will simply verify that the standards referred to in Article 5(2) have been correctly applied and will draw up a certificate of adequacy for the file,
- or submit the example of the machinery for the EC type-examination referred to in Annex VI.

3. Where the first indent of paragraph 2(c) of this Article applies, the provisions of the first sentence of paragraphs 5 and 7 of Annex VI shall also apply.

Where the second indent of paragraph 2(c) of this Article applies, the provisions of paragraphs 5, 6 and 7 of Annex VI shall also apply.

4. Where paragraph 2(a) and the first and second indents of paragraph 2(c) apply, the EC declaration of conformity shall solely state conformity with the essential requirements of the Directive.

Where paragraph 2(b) and the third indent of paragraph 2(c) apply, the EC declaration of conformity shall state conformity with the example that underwent EC type-examination.

5. Safety components shall be subject to the certification procedures applicable to machinery pursuant to paragraphs 2, 3 and 4. Furthermore, during EC type-examination, the notified body shall verify the suitability of the safety component for fulfilling the safety functions declared by the manufacturer.

6. (a) Where the machinery is subject to other Directives concerning other aspects and which also provide for the affixing of the CE marking, the latter shall indicate that the machinery is also presumed to conform to the provisions of those other Directives.

(b) However, where one or more of those Directives allow the manufacturer, during a transitional period, to choose which arrangements to apply, the CE marking shall indicate conformity only to the Directives applied by the manufacturer. In this case, particulars of the Directives applied, as published in the Official Journal of the European Communities, must be given in the documents, notices or instructions required by the directives and accompanying such machinery.

7. Where neither the manufacturer nor his authorised representative established in the Community fulfils the obligations of paragraphs 1 to 6, these obligations shall fall to any person placing the machinery or safety component on the market in the Community. The same obligations shall apply to any person assembling machinery or parts thereof or safety components of various origins or constructing machinery or safety components for his own use.

8. The obligations referred to in paragraph 7 shall not apply to persons who assemble with a machine or tractor interchangeable equipment as provided for in Article 1, provided that the parts are compatible and each of the constituent parts of the assembled machine bears the CE marking and is accompanied by the EC declaration of conformity.

ARTICLE 9

1. Member States shall notify the Commission and the other Member States of the approved bodies which they have appointed to carry out the procedures referred to in Article 8 together with the specific tasks which these bodies have been appointed to carry out and the identification numbers assigned to them beforehand by the Commission.

The Commission shall publish in the Official Journal of the European Communities a list of the notified bodies and their identification numbers and the tasks for which they have been notified. The Commission shall ensure that this list is kept up to date.

2. Member States shall apply the criteria laid down in Annex VII in assessing the bodies to be indicated in such notification. Bodies meeting the assessment criteria laid down in the relevant harmonized standards shall be presumed to fulfil those criteria.

3. A Member State which has approved a body must withdraw its notification if it finds that the body no longer meets the criteria referred to in

Annex VII. It shall immediately inform the Commission and the other Member States accordingly.

CHAPTER III
CE MARKING

ARTICLE 10

1. The CE conformity marking shall consist of the initials "CE." The form of the marking to be used is shown in Annex III.
2. The CE marking shall be affixed to machinery distinctly and visibly in accordance with point 1.7.3 of Annex I.
3. The affixing of markings on the machinery which are likely to deceive third parties as to the meaning and form of the CE marking shall be prohibited. Any other marking may be affixed to the machinery provided that the visibility and legibility of the CE marking is not thereby reduced.
4. Without prejudice to Article 7:
(a) where a Member State establishes that the CE marking has been affixed unduly, the manufacturer or his authorised representative established within the Community shall be obliged to make the product conform as regards the provisions concerning the CE marking and to end the infringement under the conditions imposed by the Member State;
(b) where non-conformity continues, the Member State must take all appropriate measures to restrict or prohibit the placing on the market of the product in question or to ensure that it is withdrawn from the market in accordance with the procedure laid down in Article 7.

CHAPTER IV
FINAL PROVISIONS

ARTICLE 11

Any decision taken pursuant to this Directive which restricts the placing on the market and putting into service of machinery or a safety component shall state the exact grounds on which it is based. Such a decision shall be notified as soon as possible to the party concerned, who shall at the same time be informed of the legal remedies available to him under the laws in force in the Member State concerned and of the time limits to which such remedies are subject.

ARTICLE 12

The Commission will take the necessary steps to have information on all the relevant decisions relating to the management of this Directive made available.

ARTICLE 13

1. Member States shall communicate to the Commission the texts of the provisions of national law which they adopt in the field governed by this Directive.

2. The Commission shall, before 1 January 1994, examine the progress made in the standardisation work relating to this Directive and propose any appropriate measures.

ARTICLE 14

1. The Directives listed in Annex VIII, Part A, are hereby repealed, without prejudice to the obligations of the Member States concerning the deadlines for transposition and application of the said Directives, as set out in Annex VIII, Part B.

2. References to the repealed Directives shall be construed as references to this Directive and be read in accordance with the correlation table set out in Annex IX.

ARTICLE 15

This Directive shall enter into force on the 20th day following that of its publication in the Official Journal of the European Communities.

ARTICLE 16

This Directive is addressed to the Member States.
Done at Luxembourg, 22 June 1998.
For the European Parliament
The President
J. M. GIL-ROBLES
For the Council
The President
J. CUNNINGHAM

ANNEX I
ESSENTIAL HEALTH AND SAFETY REQUIREMENTS RELATING TO THE DESIGN AND CONSTRUCTION OF MACHINERY AND SAFETY COMPONENTS

For the purposes of this Annex "machinery" means either "machinery" or "safety component" as defined in Article 1(2).

PRELIMINARY OBSERVATIONS

1. The obligations laid down by the essential health and safety requirements apply only when the corresponding hazard exists for the machinery in question when it is used under the conditions foreseen by the manufacturer. In any event, requirements 1.1.2, 1.7.3 and 1.7.4 apply to all machinery covered by this Directive.

2. The essential health and safety requirements laid down in this Directive are mandatory. However, taking into account the state of the art, it may not be possible to meet the objectives set by them. In this case, the machinery must as far as possible be designed and constructed with the purpose of approaching those objectives.

3. The essential health and safety requirements have been grouped according to the hazards which they cover.

Machinery presents a series of hazards which may be indicated under more than one heading in this Annex.

The manufacturer is under an obligation to assess the hazards in order to identify all of those which apply to his machine; he must then design and construct it taking account of his assessment.

1. ESSENTIAL HEALTH AND SAFETY REQUIREMENTS

1.1 General remarks

1.1.1 Definitions

For the purpose of this Directive:

1. "danger zone" means any zone within and/or around machinery in which an exposed person is subject to a risk to his health or safety;

2. "exposed person" means any person wholly or partially in a danger zone;

3. "operator" means the person or persons given the task of installing, operating, adjusting, maintaining, cleaning, repairing or transporting machinery.

1.1.2 Principles of safety integration

(a) Machinery must be so constructed that it is fitted for its function, and can be adjusted and maintained without putting persons at risk when these operations are carried out under the conditions foreseen by the manufacturer.

The aim of measures taken must be to eliminated any risk of accident throughout the foreseeable lifetime of the machinery, including the phases of assembly and dismantling, even where risks of accident arise from foreseable abnormal situations.

(b) In selecting the most appropriate methods, the manufacturer must apply the following principles, in the order given:

- eliminate or reduce risks as far as possible (inherently safe machinery design and construction),
- take the necessary protection measures in relation to risks that cannot be eliminated,
- inform users of the residual risks due to any shortcomings of the protection measures adopted, indicate whether any particular

training is required and specify any need to provide personal protection equipment.

(c) When designing and constructing machinery, and when drafting the instructions, the manufacturer must envisage not only the normal use of the machinery but also uses which could reasonably be expected.

The machinery must be designed to prevent abnormal use if such use would engender a risk. In other cases the instructions must draw the user's attention to ways-which experience has shown might occur-in which the machinery should not be used.

(d) Under the intended conditions of use, the discomfort, fatigue and psychological stress faced by the operator must be reduced to the minimum possible taking ergonomic principles into account.

(e) When designing and constructing machinery, the manufacturer must take account of the constraints to which the operator is subject as a result of the necessary or foreseeable use of personal protection equipment (such as footwear, gloves, etc.).

(f) Machinery must be supplied with all the essential special equipment and accessories to enable it to be adjusted, maintained and used without risk.

1.1.3 Materials and products

The materials used to construct machinery or products used and created during its use must not endanger exposed persons' safety or health.

In particular, where fluids are used, machinery must be designed and constructed for use without risks due to filling, use, recovery or draining.

1.1.4 Lighting

The manufacturer must supply integral lighting suitable for the operations concerned where its lack is likely to cause a risk despite ambient lighting of normal intensity.

The manufacturer must ensure that there is no area of shadow likely to cause nuisance, that there is no irritating dazzle and that there are no dangerous stroboscopic effects due to the lighting provided by the manufacturer.

Internal parts requiring frequent inspection, and adjustment and maintenance areas, must be provided with appropriate lighting.

1.1.5 Design of machinery to facilitate its handling

Machinery or each component part thereof must:

- be capable of being handled safely,
- be packaged or designed so that it can be stored safely and without damage (e.g. adequate stability, special supports, etc.).

Where the weight, size or shape of machinery or its various component parts prevents them from being moved by hand, the machinery or each component part must:

- either be fitted with attachments for lifting gear, or
- be designed so that it can be fitted with such attachments (e.g. threaded holes), or
- be shaped in such a way that standard lifting gear can easily be attached.

Where machinery or one of its component parts is to be moved by hand, it must:

- either be easily movable, or
- be equipped for picking up (e.g. hand-grips, etc.) and moving in complete safety.

Special arrangements must be made for the handling of tools and/or machinery parts, even if lightweight, which could be dangerous (shape, material, etc.).

1.2 Controls

1.2.1 Safety and reliability of control systems

Control systems must be designed and constructed so that they are safe and reliable, in a way that will prevent a dangerous situation arising. Above all they must be designed and constructed in such a way that:

- they can withstand the rigours of normal use and external factors,
- errors in logic do not lead to dangerous situations.

1.2.2 Control devices

Control devices must be:

- clearly visible and identifiable and appropriately marked where necessary,
- positioned for safe operation without hesitation or loss of time, and without ambiguity,
- designed so that the movement of the control is consistent with its effect,
- located outside the danger zones, except for certain controls where necessary, such as emergency stop, console for training of robots,
- positioned so that their operation cannot cause additional risk,
- designed or protected so that the desired effect, where a risk is involved, cannot occur without an intentional operation,
- made so as to withstand foreseeable strain; particular attention must be paid to emergency stop devices liable to be subjected to considerable strain.

Where a control is designed and constructed to perform several different actions, namely where there is no one-to-one correspondence (e.g. keyboards, etc.), the action to be performed must be clearly displayed and subject to confirmation where necessary.

Controls must be so arranged that their layout, travel and resistance to operation are compatible with the action to be performed, taking account of ergonomic principles. Constraints due to the necessary or foreseeable use of personal protection equipment (such as footwear, gloves, etc.) must be taken into account.

Machinery must be fitted with indicators (dials, signals, etc.) as required for safe operation. The operator must be able to read them from the control position.

From the main control position the operator must be able to ensure that there are no exposed persons in the danger zones.

If this is impossible, the control system must be designed and constructed so that an acoustic and/ or visual warning signal is given whenever the machinery is about to start. The exposed person must have the time and the means to take rapid action to prevent the machinery starting up.

1.2.3 Starting

It must be possible to start machinery only by voluntary actuation of a control provided for the purpose.

The same requirement applies:

- when restarting the machinery after a stoppage, whatever the cause,
- when effecting a significant change in the operating conditions (e.g. speed, pressure, etc.), unless such restarting or change in operating conditions is without risk to exposed persons.

This essential requirement does not apply to the restarting of the machinery or to the change in operating conditions resulting from the normal sequence of an automatic cycle.

Where machinery has several starting controls and the operators can therefore put each other in danger, additional devices (e.g. enabling devices or selectors allowing only one part of the starting mechanism to be actuated at any one time) must be fitted to rule out such risks.

It must be possible for automated plant functioning in automatic mode to be restarted easily after a stoppage once the safety conditions have been fulfilled.

1.2.4 Stopping device

Normal stopping

Each machine must be fitted with a control whereby the machine can be brought safely to a complete stop.

Each workstation must be fitted with a control to stop some or all of the moving parts of the machinery, depending on the type of hazard, so that the machinery is rendered safe. The machinery's stop control must have priority over the start controls.

Once the machinery or its dangerous parts have stopped, the energy supply to the actuators concerned must be cut off.

Emergency stop

Each machine must be fitted with one or more emergency stop devices to enable actual or impending danger to be averted. The following exceptions apply:

- machines in which an emergency stop device would not lessen the risk, either because it would not reduce the stopping time or because it would not enable the special measures required to deal with the risk to be taken,
- hand-held portable machines and hand-guided machines.

This device must:

- have clearly identifiable, clearly visible and quickly accessible controls,
- stop the dangerous process as quickly as possible, without creating additional hazards,
- where necessary, trigger or permit the triggering of certain safeguard movements.

Once active operation of the emergency stop control has ceased following a stop command, that command must be sustained by engagement of the emergency stop device until that engagement is specifically overridden; it must not be possible to engage the device without triggering a stop command; it must be possible to disengage the device only by an appropriate operation, and disengaging the device must not restart the machinery but only permit restarting.

Complex installations

In the case of machinery or parts of machinery designed to work together, the manufacturer must so design and construct the machinery that the stop controls, including the emergency stop, can stop not only the machinery itself but also all equipment upstream and/or downstream if its continued operation can be dangerous.

1.2.5 Mode selection

The control mode selected must override all other control systems with the exception of the emergency stop.

If machinery has been designed and built to allow for its use in several control or operating modes presenting different safety levels (e.g. to allow for adjustment, maintenance, inspection, etc.), it must be fitted with a mode selector which can be locked in each position. Each position of the selector must correspond to a single operating or control mode.

The selector may be replaced by another selection method which restricts the use of certain functions of the machinery to certain categories of operator (e.g. access codes for certain numerically controlled functions, etc.).

If, for certain operations, the machinery must be able to operate with its protection devices neutralised, the mode selector must simultaneously:

- disable the automatic control mode,
- permit movements only by controls requiring sustained action,
- permit the operation of dangerous moving parts only in enhanced safety conditions (e.g. reduced speed, reduced power, step-by-step, or other adequate provision) while preventing hazards from linked sequences,
- prevent any movement liable to pose a danger by acting voluntarily or involuntarily on the machine's internal sensors.

In addition, the operator must be able to control operation of the parts he is working on at the adjustment point.

1.2.6 Failure of the power supply

The interruption, re-establishment after an interruption or fluctuation in whatever manner of the power supply to the machinery must not lead to a dangerous situation.

In particular:

- the machinery must not start unexpectedly,
- the machinery must not be prevented from stopping if the command has already been given,

- no moving part of the machinery or piece held by the machinery must fall or be ejected,
- automatic or manual stopping of the moving parts whatever they may be must be unimpeded,
- the protection devices must remain fully effective.

1.2.7 *Failure of the control circuit*

A fault in the control circuit logic, or failure of or damage to the control circuit must not lead to dangerous situations.

In particular:

- the machinery must not start unexpectedly,
- the machinery must not be prevented from stopping if the command has already been given,
- no moving part of the machinery or piece held by the machinery must fall or be ejected,
- automatic or manual stopping of the moving parts whatever they may be must be unimpeded,
- the protection devices must remain fully effective.

1.2.8 *Software*

Interactive software between the operator and the command or control system of a machine must be user-friendly.

1.3 Protection against mechanical hazards

1.3.1 *Stability*

Machinery, components and fittings thereof must be so designed and constructed that they are stable enough, under the foreseen operating conditions (if necessary taking climatic conditions into account) for use without risk of overturning, falling or unexpected movement.

If the shape of the machinery itself or its intended installation does not offer sufficient stability, appropriate means of anchorage must be incorporated and indicated in the instructions.

1.3.2 Risk of break-up during operation

The various parts of machinery and their linkages must be able to withstand the stresses to which they are subject when used as foreseen by the manufacturer.

The durability of the materials used must be adequate for the nature of the work place foreseen by the manufacturer, in particular as regards the phenomena of fatigue, ageing, corrosion and abrasion.

The manufacturer must indicate in the instructions the type and frequency of inspection and maintenance required for safety reasons. He must, where appropriate, indicate the parts subject to wear and the criteria for replacement.

Where a risk of rupture or disintegration remains despite the measures taken (e.g. as with grinding wheels) the moving parts must be mounted and positioned in such a way that in case of rupture their fragments will be contained.

Both rigid and flexible pipes carrying fluids, particularly those under high pressure, must be able to withstand the foreseen internal and external stresses and must be firmly attached and/or protected against all manner of external stresses and strains; precautions must be taken to ensure that no risk is posed by a rupture (sudden movement, high-pressure jets, etc.).

Where the material to be processed is fed to the tool automatically, the following conditions must be fulfilled to avoid risks to the persons exposed (e.g. tool breakage):

- when the workpiece comes into contact with the tool the latter must have attained its normal working conditions,
- when the tool starts and/or stops (intentionally or accidentally) the feed movement and the tool movement must be coordinated.

1.3.3 Risks due to falling or ejected objects

Precautions must be taken to prevent risks from falling or ejected objects (e.g. workpieces, tools, cuttings, fragments, waste, etc.).

1.3.4 Risks due to surfaces, edges or angles

In so far as their purpose allows, accessible parts of the machinery must have no sharp edges, no sharp angles, and no rough surfaces likely to cause injury.

1.3.5 Risks related to combined machinery

Where the machinery is intended to carry out several different operations with the manual removal of the piece between each operation (combined machinery), it must be designed and constructed in such a way as to enable each element to be used separately without the other elements constituting a danger or risk for the exposed person.

For this purpose, it must be possible to start and stop separately any elements that are not protected.

1.3.6 Risks relating to variations in the rotational speed of tools

When the machine is designed to perform operations under different conditions of use (e.g. different speeds or energy supply), it must be designed and constructed in such a way that selection and adjustment of these conditions can be carried out safely and reliably.

1.3.7 Prevention of risks related to moving parts

The moving parts of machinery must be designed, built and laid out to avoid hazards or, where hazards persist, fixed with guards or protective devices in such a way as to prevent all risk of contact which could lead to accidents.

All necessary steps must be taken to prevent accidental blockage of moving parts involved in the work. In cases where, despite the precautions taken, a blockage is likely to occur, specific protection devices or tools, the instruction handbook and possibly a sign on the machinery should be provided by the manufacturer to enable the equipment to be safely unblocked.

1.3.8 Choice of protection against risks related to moving parts

Guards or protection devices used to protect against the risks related to moving parts must be selected on the basis of the type of risk. The following guidelines must be used to help make the choice.

A. Moving transmission parts

Guards designed to protect exposed persons against the risks associated with moving transmission parts (such as pulleys, belts, gears, rack and pinions, shafts, etc.) must be:

- either fixed, complying with requirements 1.4.1 and 1.4.2.1, or
- movable, complying with requirements 1.4.1 and 1.4.2.2.A.

Movable guards should be used where frequent access is foreseen.

B. Moving parts directly involved in the process

Guards or protection devices designed to protect exposed persons against the risks associated with moving parts contributing to the work (such as cutting tools, moving parts of presses, cylinders, parts in the process of being machined, etc.) must be:

- wherever possible fixed guards complying with requirements 1.4.1 and 1.4.2.1,
- otherwise, movable guards complying with requirements 1.4.1 and 1.4.2.2.B or protection devices such as sensing devices (e.g. non-material barriers, sensor mats), remote-hold protection devices (e.g. two-hand controls), or protection devices intended automatically to prevent all or part of the operator's body from encroaching on the danger zone in accordance with requirements 1.4.1 and 1.4.3.

However, when certain moving parts directly involved in the process cannot be made completely or partially inaccessible during operation owing to operations requiring nearby operator intervention, where technically possible such parts must be fitted with:

- fixed guards, complying with requirements 1.4.1 and 1.4.2.1 preventing access to those sections of the parts that are not used in the work,
- adjustable guards, complying with requirements 1.4.1 and 1.4.2.3 restricting access to those sections of the moving parts that are strictly for the work.

1.4 Required characteristics of guards and protection devices

1.4.1 General requirements

Guards and protection devices must:

- be of robust construction,
- not give rise to any additional risk,
- not be easy to by-pass or render non-operational,
- be located at an adequate distance from the danger zone,
- cause minimum obstruction to the view of the production process,
- enable essential work to be carried out on installation and/or replacement of tools and also for maintenance by restricting access only to the area where the work has to be done, if possible without the guard or protection device having to be dismantled.

1.4.2 Special requirements for guards

1.4.2.1 Fixed guards

Fixed guards must be securely held in place.

They must be fixed by systems that can be opened only with tools.

Where possible, guards must be unable to remain in place without their fixings.

1.4.2.2 Movable guards

A. Type A movable guards must:

- as far as possible remain fixed to the machinery when open,
- be associated with a locking device to prevent moving parts starting up as long as these parts can be accessed and to give a stop command whenever they are no longer closed.

B. Type B movable guards must be designed and incorporated into the control system so that:

- moving parts cannot start up while they are within the operator's reach,
- the exposed person cannot reach moving parts once they have started up,
- they can be adjusted only by means of an intentional action, such as the use of a tool, key, etc.,
- the absence or failure of one of their components prevents starting or stops the moving parts,

- protection against any risk of ejection is proved by means of an appropriate barrier.

1.4.2.3 Adjustable guards restricting access

Adjustable guards restricting access to those areas of the moving parts strictly necessary for the work must:

- be adjustable manually or automatically according to the type of work involved,
- be readily adjustable without the use of tools,
- reduce as far as possible the risk of ejection.

1.4.3 Special requirements for protection devices

Protection devices must be designed and incorporated into the control system so that:

- moving parts cannot start up while they are within the operator's reach,
- the exposed person cannot reach moving parts once they have started up, they can be adjusted only by means of an intentional action, such as the use of a took, key, etc.,
- the absence or failure of one of their components prevents starting or stops the moving parts.

1.5 Protection against other hazards

1.5.1 Electricity supply

Whereas machinery has an electricity supply it must be designed, constructed and equipped so that all hazards of an electrical nature are or can be prevented.

The specific rules in force relating to electrical equipment designed for use within certain voltage limits must apply to machinery which is subject to those limits.

1.5.2 Static electricity

Machinery must be so designed and constructed as to prevent or limit the build-up of potentially dangerous electrostatic charges and/or be fitted with a discharging system.

1.5.3 Energy supply other than electricity

Where machinery is powered by an energy other than electricity (e.g. hydraulic, pneumatic or thermal energy, etc.), it must be so designed, constructed and equipped as to avoid all potential hazards associated with these types of energy.

1.5.4 Errors of fitting

Errors, likely to be made when fitting or refitting certain parts which could be a source of risk must be made impossible by the design of such parts or, failing this, by information given on the parts themselves and/or the housings. The same information must be given on moving parts and/or their housings where the direction of movement must be known to avoid a risk. Any further information that may be necessary must be given in the instructions.

Where a faulty connection can be the source of risk, incorrect fluid connections, including electrical conductors, must be made impossible by the design or, failing this, by information given on the pipes, cables, etc. and/or connector blocks.

1.5.5 Extreme temperatures

Steps must be taken to eliminate any risk of injury caused by contact with or proximity to machinery parts or materials at high or very low temperatures.

The risk of hot or very cold material being ejected should be assessed. Where this risk exists, the necessary steps must be taken to prevent it or, if this is not technically possible, to render it non-dangerous.

1.5.6 Fire

Machinery must be designed and constructed to avoid all risk of fire or overheating posed by the machinery itself or by gases, liquids, dust, vapours or other substances produced or used by the machinery.

1.5.7 Explosion

Machinery must be designed and constructed to avoid any risk of explosion posed by the machinery itself or by gases, liquids, dust, vapors or other substances produced or used by the machinery.

To that end the manufacturer must take steps to:

- avoid a dangerous concentration of products,
- prevent combustion of the potentially explosive atmosphere,
- minimize any explosion which may occur so that it does not endanger the surroundings.

The same precautions must be taken if the manufacturer foresees the use of the machinery in a potentially explosive atmosphere.

Electrical equipment forming part of the machinery must conform, as far as the risk from explosion is concerned, to the provision of the specific Directives in force.

1.5.8 Noise

Machinery must be so designed and constructed that risks resulting from the emission of airborne noise are reduced to the lowest level taking account of technical progress and the availability of means of reducing noise, in particular at source.

1.5.9 Vibration

Machinery must be so designed and constructed that risks resulting from vibrations produced by the machinery are reduced to the lowest level, taking account of technical progress and the availability of means of reducing vibration, in particular at source.

1.5.10 Radiation

Machinery must be so designed and constructed that any emission of radiation is limited to the extent necessary for its operation and that the effects on exposed persons are non-existent or reduced to non-dangerous proportions.

1.5.11 External radiation

Machinery must be so designed and constructed that external radiation does not interfere with its operation.

1.5.12 Laser equipment

Where laser equipment is used, the following provisions should be taken into account:

- laser equipment on machinery must be designed and constructed so as to prevent any accidental radiation,
- laser equipment on machinery must be protected so that effective radiation, radiation produced by reflection or diffusion and secondary radiation do not damage health,
- optical equipment for the observation or adjustment of laser equipment on machinery must be such that no health risk is created by the laser rays.

1.5.13 Emissions of dust, gases, etc.

Machinery must be so designed, constructed and/or equipped that risks due to gases, liquids, dust, vapours and other waste materials which it produces can be avoided.

Where a hazard exists, the machinery must be so equipped that the said substances can be contained and/or evacuated.

Where machinery is not enclosed during normal operation, the devices for containment and/or evacuation must be situated as close as possible to the source emission.

1.5.14 Risk of being trapped in a machine

Machinery must be designed, constructed or fitted with a means of preventing an exposed person from being enclosed within it or, if that is impossible, with a means of summoning help.

1.5.15 Risk of slipping, tripping or falling

Parts of the machinery where persons are liable to move about or stand must be designed and constructed to prevent persons slipping, tripping or falling on or off these parts.

1.6 Maintenance

1.6.1 Machinery maintenance

Adjustment, lubrication and maintenance points must be located outside danger zones. It must be possible to carry out adjustment, maintenance, repair, cleaning and servicing operations while machinery is at a standstill.

If one or more of the above conditions cannot be satisfied for technical reasons, these operations must be possible without risk (see 1.2.5).

In the case of automated machinery and, where necessary, other machinery, the manufacturer must make provision for a connecting device for mounting diagnostic fault-finding equipment.

Automated machine components which have to be changed frequently, in particular for a change in manufacture or where they are liable to wear or likely to deteriorate following an accident, must be capable of being removed and replaced easily and in safety. Access to the components must enable these tasks to be carried out with the necessary technical means (tools, measuring instruments, etc.) in accordance with an operating method specified by the manufacturer.

1.6.2 Access to operating position and servicing points

The manufacturer must provide means of access (stairs, ladders, catwalks, etc.) to allow access in safety to all areas used for production, adjustment and maintenance operations.

1.6.3 Isolation of energy sources

All machinery must be fitted with means to isolate it from all energy sources. Such isolators must be clearly identified. They must be capable of being locked if reconnection could endanger exposed persons. In the case of machinery supplied with electricity through a plug capable of being plugged into a circuit, separation of the plug is sufficient.

The isolator must be capable of being locked also where an operator is unable, from any of the points to which he has access, to check that the energy is still cut off.

After the energy is cut off, it must be possible to dissipate normally any energy remaining or stored in the circuits of the machinery without risk to exposed persons.

As an exception to the above requirements, certain circuits may remain connected to their energy sources in order, for example, to hold parts, protect information, light interiors, etc. In this case, special steps must be taken to ensure operator safety.

1.6.4 Operator intervention

Machinery must be so designed, constructed and equipped that the need for operator intervention is limited.

If operator intervention cannot be avoided, it must be possible to carry it out easily and in safety.

1.6.5 Cleaning of internal parts

The machinery must be designed and constructed in such a way that it is possible to clean internal parts which have contained dangerous substances or preparations without entering them; any necessary unblocking must also be possible from the outside. If it is absolutely impossible to avoid entering the machinery, the manufacturer must take steps during its construction to allow cleaning to take place with the minimum of danger.

1.7 Indicators

1.7.0 Information devices

The information needed to control machinery must be unambiguous and easily understood.

It must not be excessive to the extent of overloading the operator.

Where the health and safety of exposed persons may be endangered by a fault in the operation of unsupervised machinery, the machinery must be equipped to give an appropriate acoustic or light signal as a warning.

1.7.1 Warning devices

Where machinery is equipped with warning devices (such as signals, etc.), these must be unambiguous and easily perceived.

The operator must have facilities to check the operation of such warning devices at all times.

The requirements of the specific Directives concerning colours and safety signals must be complied with.

1.7.2 Warning of residual risks

Where risks remain despite all the measures adopted or in the case of potential risks which are not evident (e.g. electrical cabinets, radioactive sources, bleeding of a hydraulic circuit, hazard in an unseen area, etc.), the manufacturer must provide warnings.

Such warnings should preferably use readily understandable pictograms and/or be drawn up in one of the languages of the country in which the machinery is to be used, accompanied, on request, by the languages understood by the operators.

1.7.3 Marking

All machinery must be marked legibly and indelibly with the following minimum particulars:

- name and address of the manufacturer,
- the CE marking (see Annex III),
- designation of series or type,
- serial number, if any,
- the year of construction.

Furthermore, where the manufacturer constructs machinery intended for use in a potentially explosive atmosphere, this must be indicated on the machinery.

Machinery must also bear full information relevant to its type and essential to its safe use (e.g. maximum speed of certain rotating parts, maximum diameter of tools to be fitted, mass, etc.).

Where a machine part must be handled during use with lifting equipment, its mass must be indicated legibly, indelibly and unambiguously.

The interchangeable equipment referred to in the third indent of Article 1(2)(a), must bear the same information.

1.7.4 Instructions

(a) All machinery must be accompanied by instructions including at least the following:

- a repeat of the information with which the machinery is marked, except the serial number (see 1.7.3) together with any appropriate additional information to facilitate maintenance (e.g. addresses of the importer, repairers, etc.),
- foreseen use of the machinery within the meaning of 1.1.2(c),
- workstation(s) likely to be occupied by operators,
- instructions for safe:
    - putting into service,
    - use,
    - handling, giving the mass of the machinery and its various parts where they are regularly to be transported separately,
    - assembly, dismantling,
    - adjustment,
    - maintenance (servicing and repair),
- where necessary, training instructions,
- where necessary, the essential characteristics of tools which may be fitted to the machinery.

Where necessary, the instructions should draw attention to ways in which the machinery should not be used.

(b) The instructions must be drawn up in one of the Community languages by the manufacturer or his authorised representative established in the Community. On being put into service, all machinery must be accompanied by a translation of the instructions in the language or languages of the country in which the machinery is to be used any by the instructions in the original language. This translation must be done either by the manufacturer or his authorised representative established in the Community or by the person introducing the machinery into the language area in question. By way of derogation from this requirement, the maintenance instructions for use by specialised personnel employed by the manufacturer or his authorised representative established in the Community may be drawn up in only one of the Community languages understood by that personnel.

(c) The instructions must contain the drawings and diagrams necessary for putting into service, maintenance, inspection, checking of correct operation and, where appropriate, repair of the machinery, and all useful instructions in particular with regard to safety.

(d) Any literature describing the machinery must not contradict the instructions as regards safety aspects. The technical documentation describing the machinery must give information regarding the airborne noise emissions referred to in (f) and, in the case of hand-held and/or hand-guided machinery, information regarding vibration as referred to in 2.2.

(e) Where necessary, the instructions must give the requirements relating to installation and assembly for reducing noise or vibration (e.g. use of dampers, type and mass of foundation block, etc.).

(f) The instructions must give the following information concerning airborne noise emissions by the machinery, either the actual value or a value established on the basis of measurements made on identical machinery:

- equivalent continuous A-weighted sound pressure level at workstations, where this exceeds 70 dB(A); where this level does not exceed 70 dB(A), this fact must be indicated,
- peak C-weighted instantaneous sound pressure value at workstations, where this exceeds 63 Pa (130 dB in relation to 20 µPa),
- sound power level emitted by the machinery where the equivalent continuous A-weighted sound pressure level at workstations exceeds 85 dB(A).

In the case of very large machinery, instead of the sound power level, the equivalent continuous sound pressure levels at specified positions around the machinery may be indicated.

Where the harmonized standards are not applied, sound levels must be measured using the most appropriate method for the machinery.

The manufacturer must indicate the operating conditions of the machinery during measurement and what methods have been used for the measurement.

Where the workstation(s) are undefined or cannot be defined, sound pressure levels must be measured at a distance of 1 metre from the surface of the machinery and at a height of 1.60 metres from the floor or access platform. The position and value of the maximum sound pressure must be indicated.

(g) If the manufacturer foresees that the machinery will be used in a potentially explosive atmosphere, the instructions must give all the necessary information.

(h) In the case of machinery which may also be intended for use by non-professional operators, the wording and layout of the instructions for use, whilst respecting the other essential requirements mentioned above, must take into account the level of general education and acumen that can reasonably be expected from such operators.

2. ESSENTIAL HEALTH AND SAFETY REQUIREMENTS FOR CERTAIN CATEGORIES OF MACHINERY

2.1 Agri-foodstuffs machinery

Where machinery is intended to prepare and process foodstuffs (e.g. cooking, refrigeration, thawing, washing, handling, packaging, storage, transport or distribution), it must be so designed and constructed as to avoid any risk of infection, sickness or contagion and the following hygiene rules must be observed:

(a) materials in contact, or intended to come into contact, with the foodstuffs must satisfy the conditions set down in the relevant Directives. The machinery must be so designed and constructed that these materials can be clean before each use;

(b) all surfaces including their joinings must be smooth, and must have neither ridges nor crevices which could harbour organic materials;

(c) assemblies must be designed in such a way as to reduce projections, edges and recesses to a minimum. They should preferably be made by welding or continuous bonding. Screws, screwheads and rivets may not be used except where technically unavoidable;

(d) all surfaces in contact with the foodstuffs must be easily cleaned and disinfected, where possible after removing easily dismantled parts. The inside surfaces must have curves of a radius sufficient to allow thorough cleaning;

(e) liquid deriving from foodstuffs as well as cleaning, disinfecting and rinsing fluids should be able to be discharged from the machine without impediment (possible in a "clean" position);

(f) machinery must be so designed and constructed as to prevent any liquids or living creatures, in particular insects, entering, or any organic matter accumulating in areas that cannot be cleaned (e.g. for machinery not mounted on feet or casters, by placing a seal between the machinery and its base, by the use of sealed units, etc.);

(g) machinery must be so designed and constructed that no ancillary substances (e.g. lubricants, etc.) can come into contact with foodstuffs. Where necessary, machinery must be designed and constructed so that continuing compliance with this requirement can be checked.

Instructions

In addition to the information required in section 1, the instructions must indicate recommended products and methods for cleaning, disinfecting and rinsing (not only for easily accessible areas but also where areas to which access is impossible or inadvisable, such as piping, have to be cleaned in situ).

2.2 Portable hand-held and/or hand-guided machinery

Portable hand-held and/or hand-guided machinery must conform to the following essential health and safety requirements:

- according to the type of machinery, it must have a supporting surface of sufficient size and have a sufficient number of handles and supports of an appropriate size and arranged to ensure the stability of the machinery under the operating conditions foreseen by the manufacturer,
- except where technically impossible or where there is an independent control, in the case of handles which cannot be released in complete safety, it must be fitted with start and stop controls arranged in such a way that the operator can operate them without releasing the handles,
- it must be designed, constructed or equipped to eliminate the risks of accidental starting and/or continued operation after the operator has released the handles. Equivalent steps must be taken if this requirement is not technically feasible,
- portable hand-held machinery must be designed and constructed to allow, where necessary, a visual check of the contact of the tool with the material being processed.

Instructions

The instructions must give the following information concerning vibrations transmitted by hand-held and hand-guided machinery:

- the weighted root mean square acceleration value to which the arms are subjected, if it exceeds 2.5 m/s^2 as determined by the appropriate test code. Where the acceleration does not exceed 2.5 m/s^2, this must be mentioned.

If there is no applicable test code, the manufacturer must indicate the measurement methods and conditions under which measurements were made.

2.3 Machinery for working wood and analogous materials

Machinery for working wood and machinery for working materials with physical and technological characteristics similar to those of wood, such as cork, bone, hardened rubber, hardened plastic material and other similar stiff material must conform to the following essential health and safety requirements:

(a) the machinery must be designed, constructed or equipped so that the piece being machined can be placed and guided in safety; where the piece is hand-held on a work-bench the latter must be sufficiently stable during the work and must not impede the movement of the piece;

(b) where the machinery is likely to be used in conditions involving the risk of ejection of pieces of wood, it must be designed, constructed, or equipped to eliminate this ejection, or, if this is not the case, so that the ejection does not engender risks for the operator and/or exposed persons;

(c) the machinery must be equipped with an automatic brake that stops the tool in a sufficiently short time if there is a risk of contact with the tool whilst it runs down;

(d) where the tool is incorporated into a non-fully automated machine, the latter must be so designed and constructed as to eliminate or reduce the risk of serious accidental injury, for example by using cylindrical cutter clocks restricting depth of cut, etc.

3. ESSENTIAL HEALTH AND SAFETY REQUIREMENTS TO OFFSET THE PARTICULAR HAZARDS DUE TO THE MOBILITY OF MACHINERY

Machinery presenting hazards due to mobility must be designed and constructed to meet the requirements set out below.

Risks due to mobility always exist in the case of machinery which is self-propelled, towed or pushed or carried by other machinery or tractors, is operated in working areas and whose operation requires either mobility while working, be it continuous or semi-continuous movement, between a succession of fixed working positions.

Risks due to mobility may also exist in the case of machinery operated without being moved, but equipped in such a way as to enable it to be moved more easily from one place to another machinery fitted with wheels, rollers, runners, etc. or placed on gantries, trolleys, etc.).

In order to verify that rotary cultivators and power harrows do not present unacceptable risks to the exposed persons, the manufacturer or his authorised representative established within the Community must, for each type of machinery concerned, perform the appropriate tests or have such tests performed.

3.1 General

3.1.1 Definition

"Driver" means an operator responsible for the movement of machinery. The driver may be transported by the machinery or may be on foot, accompanying the machinery, or may be guiding the machinery by remote control (cables, radio, etc.).

3.1.2 Lighting

If intended by the manufacturer to be used in dark places, self-propelled machinery must be fitted with a lighting device appropriate to the work to be carried out, without prejudice to any other regulations applicable (road traffic regulations, navigation rules, etc.).

3.1.3 Design of machinery to facilitate its handling

During the handling of the machine and/or its parts, there must be no possibility of sudden movements or of hazards due to instability as long as the machine and/or its parts are handled in accordance with the manufacturer's instructions.

3.2 Work stations

3.2.1 Driving position

The driving position must be designed with due regard to ergonomic principles. There may be two or more driving positions and, in such cases, each driving position must be provided with all the requisite controls. Where there is more than one driving position, the machinery must be designed so that the use of one of them precludes the use of the others, except in emergency stops. Visibility from the driving position must be such that the driver can in complete safety for himself and the exposed persons, operate the machinery and its tools in their intended conditions of use. Where necessary, appropriate devices must be provided to remedy hazards due to inadequate direct vision.

Machinery must be so designed and constructed that, from the driving position, there can be no risk to the driver and operators on board from inadvertent contact with the wheels or tracks.

The driving position must be designed and constructed so as to avoid any health risk due to exhaust gases and/or lack of oxygen.

The driving position of ride-on drivers must be so designed and constructed that a driver's cab may be fitted as long as there is room. In that case the cab must incorporate a place for the instructions needed for the driver and/or operators. The driving position must be fitted with an adequate cab where there is a hazard due to a dangerous environment.

Where the machine is fitted with a cab, this must be designed, constructed and/or equipped to ensure that the driver has good operating conditions and is protected against any hazards that might exist (for instance: inadequate heating and ventilation, inadequate visibility, excessive noise and vibration, falling objects, penetration by objects, rolling over, etc.). The exit must allow rapid evacuation. Moreover, an emergency exit must be provided in a direction which is different from the usual exit.

The materials used for the cab and its fittings must be fire-resistant.

3.2.2 Seating

The driving seat of any machinery must enable the driver to maintain a stable position and be designed with due regard to ergonomic principles.

The seat must be designed to reduce vibrations transmitted to the driver to the lowest level that can be reasonably achieved. The seat mountings must withstand all stresses to which they can be subjected, notably in the event of rollover. Where there is no floor beneath the driver s feet, the driver must have footrests covered with a slip-resistant material.

Where machinery is fitted with provision for a rollover protection structure, the seat must be equipped with a safety belt or equivalent device which keeps the driver in his seat without restricting any movements necessary for driving or any movements caused by the suspension.

3.2.3 Other places

If the conditions of use provide that operators other than the driver are occasionally or regularly transported by the machinery, or work on it, appropriate places must be provided which enable them to be transported or to work on it without risk, particularly the risk of falling.

Where the working conditions so permit, these work places must be equipped with seats.

Should the driving position have to be fitted with a cab, the other places must also be protected against the hazards which justified the protection of the driving position.

3.3 Controls

3.3.1 Control devices

The driver must be able to actuate all control devices required to operate the machinery from the driving position, except for functions which can be safely activated only by using control devices located away from the driving position. This refers in particular to working positions other than the driving position, for which operators other than the driver are responsible or for which the driver has to leave his driving position in order to carry out the manoeuvre in safety.

Where there are pedals they must be so designed, constructed and fitted to allow operation by the driver in safety with the minimum risk of confusion; they must have a slip-resistant surface and be easy to clean.

Where their operation can lead to hazards, notably dangerous movements, the machinery's controls, except for those with preset positions, must return to the neutral position as soon as they are released by the operator.

In the case of wheeled machinery, the steering system must be designed and constructed to reduce the force of sudden movements of the steering wheel or steering lever caused by shocks to the guide wheels.

Any control that locks the differential must be so designed and arranged that it allows the differential to be unlocked when the machinery is moving.

The last sentence of section 1.2.2 does not apply to the mobility function.

3.3.2 Starting/moving

Self-propelled machinery with a ride-on driver must be so equipped as to deter unauthorised persons from starting the engine.

Travel movements of self-propelled machinery with a ride-on driver must be possible only if the driver is at the controls.

Where, for operating purposes, machinery must be fitted with devices which exceed its normal clearance zone (e.g. stabilisers, jib, etc.), the driver must he provided with the means of checking easily, before moving the machinery, that such devices are in a particular position which allows safe movement.

This also applies to all other parts which, to allow safe movement, have to be in particular positions, locked if necessary.

Where it is technically and economically feasible, movement of the machinery must depend on safe positioning of the aforementioned parts.

It must not be possible for movement of the machinery to occur while the engine is being started.

3.3.3 Travelling function

Without prejudice to the provisions of road traffic regulations, self-propelled machinery and its trailers must meet the requirements for slowing down, stopping, braking and immobilization so as to ensure safety under all the operating, loading, speed, ground and gradient conditions allowed for by the manufacturer and corresponding to conditions encountered in normal use.

The driver must be able to slow down and stop self-propelled machinery by means of a main device. Where safety so requires in the event of a failure of the main device, or in the absence of the energy supply to actuate the main device, an emergency device with fully independent and easily accessible controls must be provided for slowing down and stopping.

Where safety so requires, a parking device must be provided to render stationary machinery immobile. This device may be combined with one of the devices referred to in the second paragraph, provided that it is purely mechanical.

Remote-controlled machinery must be designed and constructed to stop automatically if the driver loses control.

Section 1.2.4 does not apply to the travelling function.

3.3.4 Movement of pedestrian-controlled machinery

Movement of pedestrian-controlled self-propelled machinery must be possible only through sustained action on the relevant control by the driver. In particular, it must not be possible for movement to occur while the engine is being started.

The control systems for pedestrian-controlled machinery must be designed to minimize the hazards arising from inadvertent movement of the machine towards the driver. In particular:

(a) crushing;

(b) injury from rotating tools.

Also, the speed of normal travel of the machine must be compatible with the pace of a driver on foot.

In the case of machinery on which a rotary tool may be fitted, it must not be possible to activate that tool when the reversing control is engaged, except where movement of the machinery results from movement of the tool. In the latter case, the reversing speed must be such that it does not endanger the driver.

3.3.5 Control circuit failure

A failure in the power supply to the power-assisted steering, where fitted, must not prevent machinery from being steered during the time required to stop it.

3.4 Protection against mechanical hazards

3.4.1 Uncontrolled movements

When a part of a machine has been stopped, any drift away from the stopping position, for whatever reason other than action at the controls, must be such that it is not a hazard to exposed persons.

Machinery must be so designed, constructed and where appropriate placed on its mobile support so as to ensure that when moved the uncontrolled oscillations of its center of gravity do not affect its stability or exert excessive strain on its structure.

3.4.2 Risk of break-up during operation

Parts of machinery routing at high speed which, despite the measures taken, may break up or disintegrate, must be mounted and guarded in such a way that, in case of breakage, their fragments will be contained or, if that is not possible, cannot be projected towards the driving and/or operation positions.

3.4.3 Rollover

Where, in the case of self-propelled machinery with a ride-on driver and possibly ride-on operators, there is a risk of rolling over, the machinery must be designed for and be fitted with anchorage points allowing it to be equipped with a rollover protective structure (ROPS).

This structure must be such that in case of rolling over it affords the ride-on driver and where appropriate the ride-on operators an adequate deflation-limiting volume (DLV).

In order to verify that the structure complies with the requirement laid down in the second paragraph, the manufacturer or his authorised representative established within the Community must, for each type of structure concerned, perform appropriate tests or have such tests performed.

In addition, the earth-moving machinery listed below with a capacity exceeding 15 kW must be fitted with a rollover protective structure

- crawler loaders or wheel loaders,
- backhoe loaders,
- crawler tractors or wheel tractors,
- scrapers, self-loading or not,
- graders,
- articulated steer dumpers.

3.4.4 Falling objects

Where, in the case of machinery with a ride-on driver and possibly ride-on operators, there is a risk due to falling objects or material, the machinery should be designed for if its size allows, anchorage points allowing it to be equipped with a falling object protective structure (FOPS).

This structure must be such that in the case of falling objects or material, it guarantees the ride-on operators an adequate deflection-limiting volume (DLV).

In order to verify that the structure complies with the requirement laid down in the second paragraph, the manufacturer or his authorised representative established within the Community must, for each type of structure concerned, perform appropriate tests or have such tests performed.

3.4.5 Means of access

Handholds and steps must be designed, constructed and arranged in such a way that the operators use them instinctively and do not use the controls for that purpose.

3.4.6 Towing devices

All machinery used to tow or to be towed must be fitted with towing or coupling devices designed, constructed and arranged to ensure easy and safe connection and disconnection, and to prevent accidental disconnection during use.

In so far as the towbar load requires, such machinery must be equipped with a support with a bearing surface suited to the load and the ground.

3.4.7 Transmission of power between self-propelled machinery (or tractor) and recipient machinery

Transmission shafts with universal joints linking self-propelled machinery (or tractor) to the first fixed bearing of recipient machines must be guarded on the self-propelled machinery side and the recipient machinery side over the whole length of the shaft and associated universal joints.

On the side of the self-propelled machinery (or tractor), the power take-off to which the transmission shaft is attached must be guarded either by a screen fixed to the self-propelled machinery (or tractor) or by any other device offering equivalent protection.

On the towed machinery side, the input shaft must be enclosed in a protective casing fixed to the machinery.

Torque limiters or freewheels may be fitted to universal joint transmissions only on the side adjoining the driven machine. The universal-joint transmission shaft must be marked accordingly.

All towed machinery whose operation requires a transmission shaft to connect it to self-propelled machinery or a tractor must have a system for attaching the transmission shaft so that when the machinery is uncoupled the transmission shaft and its guard are not damaged by contact with the ground or part of the machinery.

The outside parts of the guard must be so designed, constructed and arranged that they cannot turn with the transmission shaft. The guard must cover the transmission shaft to the ends of the inner jaws in the case of simple universal joints and at least to the center of the outer joint or joints in the case of "wide-angle" universal joints.

Manufacturers providing means of access to working positions near to the universal joint transmission shaft must ensure that shaft guards as described in the sixth paragraph cannot be used as steps unless designed and constructed for that purpose.

3.4.8 Moving transmission parts

By way of derogation from section 1.3.8.A, in the case of internal combustion engines, removable guards preventing access to the moving parts in the engine compartment need not have locking devices if they have to be opened either by the use of a tool or key or by a control located in the driving position if the latter is in a fully enclosed cab with a lock to prevent unauthorized access.

3.5 Protection against other hazards

3.5.1 Batteries

The battery housing must be constructed and located and the battery installed so as to avoid as far as possible the chance of electrolyte being ejected on to the operator in the event of rollover and/or to avoid the accumulation of vapors in places occupied by operators.

Machinery must be so designed and constructed that the battery can be disconnected with the aid of an easily accessible device provided for that purpose.

3.5.2 Fire

Depending on the hazards anticipated by the manufacturer when in use, machinery must, where its size permits:

- either allow easily accessible fire extinguishes to be fitted,
- or be provided with built-in extinguisher systems.

3.5.3 Emissions of dust, gases, etc.

Where such hazards exist, the containment equipment provided for in section 1.5.13 may be replaced by other means, for example precipitation by water spraying.

The second and third paragraphs of section 1.5.13 do not apply where the main function of the machinery is the spraying of products.

3.6 Indications

3.6.1 Signs and warning

Machinery must have means of signalling and/or instruction plates concerning use, adjustment and maintenance, wherever necessary, to ensure the health and safety of exposed persons. They must be chosen, designed and constructed in such a way as to be clearly visible and indelible.

Without prejudice to the requirements to be observed for travelling on the public highway, machinery with a ride-on driver must have the following equipment:

- an acoustic warning device to alert exposed persons,
- a system of light signals relevant to the intended conditions of use such as stop lamps, reversing lamps and rotating beacons. The latter requirement does not apply to machinery intended solely for underground working and having no electrical power.

Remote-controlled machinery which under normal conditions of use exposes persons to the hazards of impact or crushing must be fitted with appropriate means to signal its movements or with means to protect exposed persons against such hazards. The same applies to machinery which involves, when in use, the constant repetition of a forward and backward movement on a single axis where the back of the machine is not directly visible to the driver.

Machinery must be so constructed that the warning and signalling devices cannot all be disabled unintentionally. Where this is essential for safety, such devices must be provided with the means to check that they are in good working order and their failure must be made apparent to the operator.

Where the movement of machinery or its tools is particularly hazardous, signs on the machinery must be provided to warn against approaching the machinery while it is working; the signs must be legible at a sufficient distance to ensure the safety of persons who have to be in the vicinity.

3.6.2 Marking

The minimum requirements set out in 1.7.3 must be supplemented by the following:

- nominal power expressed in kW,
- mass in kg of the most usual configuration and, where appropriate:
  - maximum drawbar pull provided for by the manufacturer at the coupling hook, in N,
  - maximum vertical load provided for by the manufacturer on the coupling hook, in N.

3.6.3 Instruction handbook

Apart from the minimum requirements set out in 1.7.4, the instruction handbook must contain the following information:

(a) regarding the vibrations emitted by the machinery, either the annual value or a figure calculated from measurements performed on identical machinery:
- the weighted root mean square acceleration value to which the arms are subjected, if it exceeds 2.5 m/s², should it not exceed 2.5 m/s², this must be mentioned.

- the weighted root mean square acceleration value to which the body (feet or posterior) is subjected, if it exceeds 0.5 m/s², should it not exceed 0.5 m/s², this must be mentioned.

Where the harmonized standards are not applied, the vibration must be measured using the most appropriate method for the machinery concerned

The manufacturer must indicate the operating conditions of the machinery during measurement and which methods were used for taking the measurements;

(b) in the case of machinery allowing several uses depending on the equipment used, manufacturers of basic machinery to which interchangeable equipment may be attached and manufacturers of the interchangeable equipment must provide the necessary information to enable the equipment to be fitted and used safely.

4. ESSENTIAL HEALTH AND SAFETY REQUIREMENTS TO OFFSET THE PARTICULAR HAZARDS DUE TO A LIFTING OPERATION

Machinery presenting hazards due to lifting operations-mainly hazards of load falls and collisions or hazards of tipping caused by a lifting operation-must be designed and constructed to meet the requirements set out below.

Risks due to a lifting operation exist particularly in the case of machinery designed to move a unit load involving a change in level during the movement. The load may consist of objects, materials or goods.

4.1 General remarks

4.1.1 Definitions

(a) "lifting accessories" means components or equipment not attached to the machine and placed between the machinery and the load or on the load in order to attach it;

(b) "separate lifting accessories" means accessories which help to make up or use a slinging device, such as eyehooks, shackles, rings, eyebolts, etc.;

(c) "guided load" means the load where the total movement is made along rigid or flexible guides, whose position is determined by fixed points;

(d) "working coefficient" means the arithmetic ratio between the load guaranteed by the manufacturer up to which a piece of equipment, an accessory or machinery is able to hold it and the maximum working load marked on the equipment, accessory or machinery respectively;

(e) "test coefficient" means the arithmetic ratio between the load used to carry out the static or dynamic tests on a piece of equipment, an accessory or machinery and the maximum working load marked on the piece of equipment, accessory or machinery;

(f) "static test" means the test during which the machinery or the lifting accessory is first inspected and subjected to a force corresponding to the maximum working load multiplied by the appropriate static test coefficient and then re-inspected once the said load has been released to ensure no damage has occurred;

(g) "dynamic test" means the test during which the machinery is operated in all its possible configurations at maximum working load with account being taken of the dynamic behavior of the machinery in order to check that the machinery) and safety features are functioning properly.

4.1.2 Protection against mechanical hazards

4.1.2.1 Risks due to lack of stability

Machinery must be so designed and constructed that the stability required in 1.3.1 is maintained both in service and out of service, including all stages of transportation, assembly and dismantling, during foreseeable component failures and also during the tests carried out in accordance with the instruction handbook.

To that end, the manufacturer or his authorised representative established within the community must use the appropriate verification methods-in particular, for self-propelled industrial trucks with lift exceeding 1.80 m, the manufacturer or his authorised representative established within the. Community must, for each type of industrial truck concerned, perform a platform stability test or similar test, or have such tests performed.

4.1.2.2 Guide rails and rail tracks

Machinery must be provided with devices which act on the guide rails or tracks to prevent derailment.

However, if derailment occurs despite such devices, or if there is a failure of a rail or of a running component, devices must be provided which prevent the equipment, component or load from falling or the machine overturning.

4.1.2.3 Mechanical strength

Machinery, lifting accessories and removable components must be capable of withstanding the stresses to which they are subjected, both in and, where applicable, out of use, under the installation and operating conditions provided for by the manufacturer, and in all relevant configurations, with due regard, where appropriate, to the effects of atmospheric factors and forces exerted by persons. This requirement must also be satisfied during transport, assembly and dismantling.

Machinery and lifting accessories must be designed and constructed so as to prevent failure from fatigue or wear, taking due account of their intended use.

The materials used must be chosen on the basis of the working environments provided for by the manufacturer, with special reference to corrosion, abrasion, impacts, cold brittleness and ageing.

The machinery and the lifting accessories must be designed and constructed to withstand the overload in the static tests without permanent deformation or patent defect. The calculation must take account of the values of the static test coefficient chosen to guarantee an adequate level of safety: that coefficient has, as a general rule, the following values:

(a) manually-operated machinery and lifting accessories 1.5;

(b) other machinery: 1.25;

Machinery must be designed and constructed to undergo, without failure, the dynamic tests carried out using the maximum working load multiplied by the dynamic test coefficient. This dynamic test coefficient is chosen so as to guarantee an adequate level of safety: the coefficient is, as a general rule, equal to 1.1.

The dynamic tests must be performed on machinery ready to be put into service under normal conditions of use. As a general rule, the tests will be performed at the nominal speeds laid down by the manufacturer. Should the control circuit of the machinery allow for a number of simultaneous movements (for example, rotation and displacement of the load), the tests must be carried out under the least favorable conditions, i.e. as a general rule, by combining the movements concerned.

4.1.2.4 Pulleys, drums, chains or ropes

Pulleys, drums and wheels must have a diameter commensurate with the size of rope or chains with which they can be fitted.

Drums and wheels must be so designed, constructed and installed that the ropes or chains with which they are equipped can wind round without falling off.

Ropes used directly for lifting or supporting the load must not include any splicing other than at their ends (splicing are tolerated in installations which are internet from their design to be modified regularly according to needs of use). Complete ropes and their endings have a working coefficient chosen so as to guarantee an adequate level of safety as a general rule, this coefficient is equal to five.

Lifting chains have a working coefficient chosen so as to guarantee an adequate level of safety; as a general rule, this coefficient is equal to four.

In order to verify that an adequate working coefficient has been attained, the manufacturer or his authorised representative established within the Community must, for each type of chain and rope used directly for lifting the load, and for the rope ends perform the appropriate tests or have such tests performed.

4.1.2.5 Separate lifting accessories

Lifting accessories must be sized with due regard to fatigue and ageing processes for a number of operating cycles consistent with their expected life-span as specified in the operating conditions for a given application.

Moreover:

(a) the working coefficient of the metallic rope/rope-end combination is chosen so as to guarantee an adequate level of safety; this coefficient is, as a general rule, equal to five. Ropes must not comprise any splices or loops other than at their ends;

(b) where chains with welded links are used, they must be of the short-link type. The working coefficient of chains of any type is chosen so as to guarantied an adequate level of safety; this coefficient is, as a general rule/ equal to four;

(c) the working coefficient for textile ropes or slings is dependent on the materials method of manufacture, dimensions and use. This coefficient is chosen so as to guarantee an adequate level of safety; it is, as a general rule, equal to seven, provided the materials used are shown to be of very good quality and the method of manufacture is appropriate to the intended use. Should this not be the case, the coefficient is, as a general rule, set at a higher level in order to secure an equivalent level of safety.

Textile ropes and slings must not include any knots, connections or splicing other than at the ends of the sling, except in the case of an endless sling;

d) all metallic components making up, or used with, a sling must have a working coefficient chosen so as to guarantee an adequate level of safety; this coefficient is, as a general rule, equal to four;

(e) the maximum working capacity of a multilegged sling is determined on the basis of the safety coefficient of the weakest leg the number of legs and a reduction factor which depends on the slinging configuration;

(f) in order to verify that an adequate working coefficient has been attained, the manufacturer or his authorised representative established within the Community must, for each type of component referred to in (a), (b), (c) and (d) perform the appropriate tests or have such tests performed.

4.1.2.6 Control of movements

Devices for controlling movements must act in such a way that the machinery on which they are installed is kept safe:

(a) machinery must be so designed or fined with devices that the amplitude of movement of its components is kept within the specified limits. The operation of such devices must, where appropriate, be preceded by a warning;

(b) where several fixed or rail-mounted machines can be manoeuvred simultaneously in the same place, with risks of collision, such machines must be so designed and constructed as to make it possible to fit systems enabling these risks to be avoided;

(c) the mechanisms of machinery must be so designed and constructed that the loads cannot creep dangerously or fall freely and unexpectedly, even in the event of partial or total failure of the power supply or when the operator stops operating the machine;

(d) it must not be possible, under normal operating conditions, to lower the load solely by friction brake, except in the case of machinery whose function requires it to operate in that way;

(e) holding devices must be so designed and constructed that inadvertent dropping of the loads is avoided.

4.1.2.7 Handling of loads

The driving position of machinery must be located in such a way as to ensure the widest possible view of trajectories of the moving parts, in order to avoid possible collisions with persons or equipment or other machinery which might be manoeuvering at the same time and liable to constitute a hazard.

Machinery with guided loads fixed in one place must be designed and constructed so as to prevent exposed persons from being hit by the load or the counter-weights.

4.1.2.8 Lightning

Machinery in need of protection against the effects of lightning while being used must be fitted with a system for conducting the resultant electrical charges to earth.

4.2 Special requirements for machinery whose power source is other than manual effort

4.2.1 Controls

4.2.1.1 Driving position

The requirements laid down in section 3.2.1 also apply to non-mobile machinery:

4.2.1.2 Seating

The requirements laid down in section 3.2.2, first and second paragraphs, and those laid down in section 3.2.3 also apply to non-mobile machinery.

4.2.1.3 Control devices

The devices controlling moments of the machinery or its equipment must return to their neutral position as soon as they are released by the operator. However, for partial or complete movements in which there is no risk of the load or the machinery colliding, the said devices may be replaced by controls authorising automatic stops at preselected levels without holding a hold-to-run control device.

4.2.1.4 Loading control

Machinery with a maximum working load of not less than 1000 kilograms or an overturning moment of not less than 40,000 N · m must be fitted with devices to warn the driver and prevent dangerous movements of the load in the event of:

- overloading the machinery:
    - either as a result of maximum working loads being exceeded, or
    - as a result of the moments due to the loads being exceeded,
- the moments conducive to overturning being exceeded as a result of the load being lifted.

4.2.2 Installation guided by cables

Cable carriers, tractors or tractor carriers must be held by counterweights or by a device allowing permanent control of the tension.

4.2.3 Risks to exposed persons. Means of access to driving position and intervention points

Machinery with guided loads and machinery whose load supports follow a clearly defined path must be equipped with devices to prevent any risks to exposed persons.

Machinery serving specific levels at which operators can gain access to the load platform in order to stack or secure the load must be designed and constructed to prevent uncontrolled movement of the load platform, in particular while being loaded or unloaded.

4.2.4 Fitness for purpose

When machinery is placed on the market or is first put into service, the manufacturer or his authorized representative established within the Community must ensure, by taking appropriate measures or having them taken, that lifting accessories and machinery which are ready for use, whether manually or power-operated, can fulfil their specified functions safely. The said measures must take into account the static and dynamic aspects of the machinery.

Where the machinery cannot be assembled in the manufacturer's premises, or in the premises of his authorized representative established within the Community, appropriate measures must be taken at the place of use. Otherwise, the measures may be taken either in the manufacturer's premises or at the place of use.

4.3 Marking

4.3.1 Chains and ropes

Each length of lifting chain, rope or webbing not forming pan of an assembly must bear a mark or, where this is not possible, a plate or irremovable ring bearing the name and address of the manufacturer or his authorised representative established in the Community and the identifying reference of the relevant certificate.

The certificate should show the information required by the harmonized standards or, should those not exist, at least the following information:

- the name of the manufacturer or his authorised representative: established within the Community,
- the address within the Community of the manufacturer or his authorised representative, as appropriate,
- a description of the chain or rope which includes:
  - its nominal size,
  - its construction,
  - the material from which it is made, and
  - any special metallurgical treatment applied to the material,
- if tested, the standard used,
- a maximum load to which the chain or rope should be subjected in service. A range of values may be given for specified applications.

4.3.2 Lifting accessories

All lifting accessories must show the following particulars:

- identification of the manufacturer,
- identification of the material (e.g. international classification) where this information is needed for dimensional compatibility,
- Identification of the maximum working load,
- CE marking.

In the case of accessories including components such as cables or ropes, on which marking is physically impossible, the particulars referred to in the first paragraph must be displayed on a plate or by some other means and securely affixed to the accessory.

The particulars must be legible and located in a place where they are not liable to disappear as a result of machining, wear, etc., or jeopardize the strength of the accessory.

4.3.3 *Machinery*

In addition to the minimum information provided for in 1.7.3, each machine must bear, legibly and indelibly, information concerning the nominal load:

(i) displayed in uncoded form and prominently on the equipment in the case of machinery which has only one possible value;

(ii) where the nominal load depends on the configuration of the machine, each driving position must be provided with a load plate indicating, preferably in diagrammatic form or by means of tables, the nominal loads for each configuration.

Machinery equipped with a load support which allows access to persons and involves a risk of falling must bear a clear and indelible warning prohibiting the lifting of persons. This warning must be visible at each place where access is possible.

4.4 Instruction handbook

4.4.1 *Lifting accessories*

Each lifting accessory or each commercially indivisible batch of lifting accessories must be accompanied with an instruction handbook setting out at least the following particulars:

- normal conditions of use,

- instructions for use, assembly and maintenance,
- the limits of use (particularly for the accessories which cannot comply with 4.1.2.6(e)).

4.4.2 Machinery

In addition to section 1.7.4, the instruction handbook must include the following information:

(a) the technical characteristics of the machinery, and in particular:

- where appropriate, a copy of the load table described in section 4.3.3(ii),
- the reactions at the supports or anchors and characteristics of the tracks,
- where appropriate, the definition and the means of installation of the ballast;

(b) the contents of the logbook, if the latter is not supplied with the machinery;

(c) advice for use, particularly to offset the lack of direct sight of the load by the operator;

(d) the necessary instructions for performing the tests before first putting into service machinery which is not assembled on the manufacturer's premises in the form in which it is to be used.

5. ESSENTIAL HEALTH AND SAFETY REQUIREMENTS FOR MACHINERY INTENDED FOR UNDERGROUND WORK

Machinery intended for underground work must be designed and constructed to meet the requirements set out below.

5.1 Risks due to lack of stability. Powered roof supports must be so designed and constructed as to maintain a given direction moving and not slip before and while they come under load and after the load has been removed. They must be equipped with anchorages for the top plates of the individual hydraulic props.

5.2 Movement. Powered roof supports must allow for unhindered movement of exposed persons.

5.3 Lighting. The requirements laid down in the third paragraph of section 1.1.4 do not apply.

5.4 Control devices. The accelerator and brake controls for the movement of machinery running on rails must be manual. The deadman's control may be foot-operated, however.

The control devices of powered roof supports must be designed and laid out so that, during displacement operations, operators are sheltered by a support in place. The control devices must be protected against any accidental release.

5.5 Stopping. Self-propelled machinery running on rails for use in underground work must be equipped with a deadman's control acting on the circuit controlling the movement of the machinery.

5.6 Fire. The second indent of 3.5.2 is mandatory in respect of machinery which comprises highly flammable parts.

The braking system of machinery meant for use in underground working must be designed and constructed so as not to produce sparks or cause fires.

Machinery with heat engines for use in underground working must be fitted only with internal combustion engines using fuel with a low vaporising pressure and which exclude any spark of electrical origin.

5.7 Emissions of dust, gases, etc. Exhaust gases from internal combustion engines must not be discharged upwards.

6. ESSENTIAL HEALTH AND SAFETY REQUIREMENTS TO OFFSET THE PARTICULAR HAZARDS DUE TO THE LIFTING OR MOVING OF PERSONS

Machinery presenting hazards due to the lifting or moving of persons must be designed and constructed to meet the requirements set out below.

6.1 General

6.1.1 Definition

For the purposes of this Chapter, "carrier" means the device by which persons are supported in order to be lifted, lowered or moved.

6.1.2 Mechanical strength

The working coefficients defined in heading 4 are inadequate for machinery intended for the lifting or moving of persons and must, as a general rule, be doubled. The floor of the carrier must be designed and constructed to offer the space and strength corresponding to the maximum number of persons and the maximum working load set by the manufacturer.

6.1.3 Loading control for types of device moved by power other than human strength

The requirements of 4.2.1.4 apply regardless of the maximum working load figure. This requirement does not apply to machinery in respect of which the manufacturer can demonstrate that there is no risk of overloading and/or overturning.

6.2 Controls

6.2.1 Where safety requirements do not impose other solutions:

The carrier must, as a general rule, be designed and constructed so that persons inside have means of controlling movements upwards and downwards and, if appropriate, of moving the carrier horizontally in relation to the machinery.

In operation, those controls must override the other devices controlling the same movement, with the exception of the emergency stop devices.

The controls for these movements must be of the maintained command type, except in the case of machinery serving specific levels.

6.2.2

If machinery for the lifting or moving of persons can be moved with the carrier in a position other than the rest position, it must be designed and constructed so that the person or persons in the carrier have the means of preventing hazards produced by the movement of the machinery.

6.2.3

Machinery for the lifting or moving of persons must be designed, constructed or equipped so that excess speeds of the carrier do not cause hazards.

6.3 Risks of persons falling from the carrier.

6.3.1

If the measures referred to in 1.1.15 are not adequate, carriers must be fitted with a sufficient number of anchorage points for the number of persons possibly using the carrier, strong enough for the attachment of personal protective equipment against the danger of falling.

6.3.2

Any trapdoors in floors or ceilings or side doors must open in a direction which obviates any risk of falling should they open unexpectedly.

6.3.3

Machinery for lifting or moving must be designed and constructed to ensure that the floor of the carrier does not tilt to an extent which creates a risk of the occupants falling, including when moving.

The floor of the carrier must be slip-resistant.

6.4 Risks of the carrier falling or overturning

6.4.1

Machinery for the lifting or moving of persons must be designed and constructed to prevent the carrier falling or overturning.

6.4.2

Acceleration and braking of the carrier or carrying vehicle, under the control of the operator or triggered by a safety device and under the maximum load and speed conditions laid down by the manufacturer, must not cause any danger to exposed persons.

6.5 Markings. Where necessary to ensure safety, the carrier must bear the relevant essential information.

ANNEX II

A. Contents of the EC declaration of conformity for machinery[1]

The EC declaration of conformity must contain the following particulars:

- name and address of the manufacturer or his authorised representative established in the Community,[2]
- description of the machinery,[3]
- all relevant provisions complied with by the machinery,
- where appropriate, name and address of the notified body and number of the EC type-examination certification,
- where appropriate, the name and address of the notified body to which the file has been forwarded in accordance with the first indent of Article 8(2)(c),
- where appropriate, the name and address of the notified body which has carried out the verification referred to in the second indent of Article 8(2)(c),
- where appropriate, a reference to the harmonized standards,
- where appropriate, the national technical standards and specifications used,

---

[1] This declaration must be drawn up in the same language as the original instructions (see Annex I, section 1.7.4(b)) and must be either typewritten or handwritten in block capitals. It must be accompanied by a translation in one of the official languages of the country in which the machinery is to be used. This translation must be done in accordance with the same conditions as for the translation of the instructions.

[2] Business name and full address; authorized representatives must also give the business name and address of the manufacturer.

[3] Description of the machinery (make, type, serial number, etc.).

- identification of the person empowered to sign on behalf of the manufacturer or his authorised representatives.

B. Contents of the declaration by the manufacturer or his authorised representatives established in the Community (Article 4(2))

The manufacturer's declaration referred to in Article 4(2) must contain the following particulars:

- name and address of the manufacturer or the authorized representative,
- description of the machinery or machinery parts,
- where appropriate, the name and address of the notified body and the number of the EC type-examination certificate,
- where appropriate, the name and address of the notified body to which the file has been forwarded in accordance with the first indent of Article 8(2)(c),
- where appropriate, the name and address of the notified body which has carried out the verification referred to in the second indent of Article 8(2)(c),
- where appropriate, a reference to the harmonized standards,
- a statement that the machinery must not be put into service under the machinery into which it is to be incorporated has been declared in conformity with the provisions of the Directive,
- identification of the person signing.

C. Contents of the EC declaration of conformity for safety components placed on the market separately[4]

The EC declaration of conformity must contain the following particulars:

- name and address of the manufacturer or his authorised representative established in the Community[5]
- description of the safety component[6]

---

[4] See note 1.
[5] See note 2.
[6] Description of the safety component (make, type, serial number, if any, etc.).

- safety function fulfilled by the safety component, if not obvious from the description,
- where appropriate, the name and address of the notified body and the number of the EC type-examination certificate,
- where appropriate, the name and address of the notified body to which the file was forwarded in accordance with the first indent of Article 8(2)(c),
- where appropriate, the name and address of the notified body which carried out the verification referred to in the second indent of Article 8(2)(c),
- where appropriate, a reference to the harmonized standards,
- where appropriate, the national technical standards and specifications used,
- identification of the person empowered to sign on behalf of the manufacturer or his authorised representative established in the Community.

ANNEX III
CE CONFORMITY MARKING

- The CE conformity marking shall consist of the initials "CE" taking the following form:

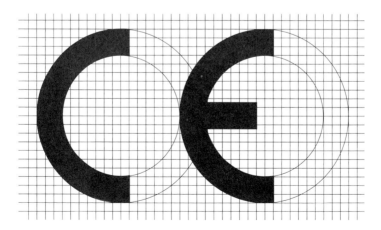

- if the CE marking is reduced or enlarged the proportions given in the above drawing must be respected,

- the various components of the CE marking must have substantially the same vertical dimension, which may not be less than 5 mm. This minimum dimension may be waived for small-scale machinery.

ANNEX IV
TYPES OF MACHINERY AND SAFETY COMPONENTS FOR WHICH THE PROCEDURE REFERRED TO IN ARTICLE 8(2)(B) AND (C) MUST BE APPLIED

A. Machinery

1. Circular saws (single or multi-blade) for working with wood and analogous materials or for working with meat and analogous materials.

1.1 Sawing machines with fixed tool during operation, having a fixed bed with manual feed of the workpiece or with a demountable power feed.

1.2 Sawing machines with fixed tool during operation, having a manually operated reciprocating saw-bench or carriage.

1.3 Sawing machines with fixed tool during operation, having a built-in mechanical feed device for the work-pieces, with manual loading and/or unloading.

1.4 Sawing machines with movable tool during operation, with a mechanical feed device and manual loading and/or unloading.

2. Hand-fed surface planing machines for woodworking.

3. Thicknessers for one-side dressing with manual loading and/or unloading for woodworking.

4. Band-saws with a fixed or mobile bed and band-saws with a mobile carriage, with manual loading and/or unloading, for working with wood and analogous materials or for working with meat and analogous materials.

5. Combined machines of the types referred to in 1 to 4 and 7 for working with wood and analogous materials.

6. Hand-fed tenoning machines with several tool holders for woodworking.

7. Hand-fed vertical spindle moulding machines for working with wood and analogous materials.

8. Portable chainsaws for woodworking.

9. Presses, including press-brakes, for the cold working of metals, with manual loading and/or unloading, whose movable working parts may have a travel exceeding 6 mm and a speed exceeding 30 mm/s.

10. Injection or compression plastics-moulding machines with manual loading or unloading.

11. Injection or compression rubber-moulding machines with manual loading or unloading.

12. Machinery for underground working of the following types:

- machinery on rails: locomotives and brake-vans,
- hydraulic-powered roof supports,
- internal combustion engines to be fitted to machinery for underground working.

13. Manually-loaded trucks for the collection of household refuse incorporating a compression mechanism.

14. Guards and detachable transmission shafts with universal joints as described in section 3.4.7.

15. Vehicles servicing lifts.

16. Devices for the lifting of persons involving a risk of falling from a vertical height of more than three metres.

17. Machines for the manufacture of pyrotechnics.

B. *Safety components*

1. Electro-sensitive devices designed specifically to detect persons in order to ensure their safety (non-material barriers, sensor mats, electro-magnetic detectors, etc.).

2. Logic units which ensure the safety functions of bimanual controls.

3. Automatic movable screens to protect the presses referred to in 9, 10 and 11.

4. Roll-over protection structures (**ROPS**).

5. Falling-object protective structures (**FOPS**).

ANNEX V
EC DECLARATION OF CONFORMITY

For the purposes of this Annex, "machinery" means either "machinery" or "safety component" as defined in Article 1(2).

1. The EC declaration of conformity is the procedure by which the manufacturer, or his authorised representative established in the Community declares that the machinery being placed on the market complies with all the essential health and safety requirements applying to it.

2. Signature of the EC declaration of conformity authorises the manufacturer, or his authorized representative in the Community, to affix the CE marking to the machinery.

3. Before drawing up the EC declaration of conformity, the manufacturer, or his authorised representative in the Community, shall have ensured and be able to guarantee that the documentation listed below is and will remain available on his premises for any inspection purposes:

(a) a technical construction file comprising:

- an overall drawing of the machinery together with drawings of the control circuits,
- full detailed drawings, accompanied by any calculation notes, test results, etc., required to check the conformity of the machinery with the essential health and safety requirements,
- a list of:
    - the essential requirements of this Directive,
    - standards, and
    - other technical specifications, which were used when the machinery was designed,
- a description of methods adopted to eliminate hazards presented by the machinery,
- if he so desires, any technical report or certificate obtained from a competent body or laboratory,[1]
- if he declares conformity with a harmonized standard which provides therefor, any technical report giving the results of tests carried out at his choice either by himself or by a competent body or laboratory,[2]

---

[1] A body or laboratory is presumed competent if it meets the assessment criteria laid down in the relevant harmonized standards.
[2] See note 1.

- a copy of the instructions for the machinery;

(b) for series manufacture, the internal measures that will be implemented to ensure that the machinery remains in conformity with the provisions of the Directive.

The manufacturer must carry out necessary research or tests on components, fittings or the completed machine to determine whether by its design or construction, the machine is capable of being erected and put into service safely.

Failure to present the documentation in response to a duly substantiated request by the competent national authorities may constitute sufficient grounds for doubting the presumption of conformity with the requirements of the Directive.

4. (a) The documentation referred to in 3 above need not permanently exist in a material manner but it must be possible to assemble it and make it available within a period of time commensurate with its importance.

It does not have to include detailed plans or any other specific information as regards the sub-assemblies used for the manufacture of the machinery unless a knowledge of them is essential for verification of conformity with the basic safety requirements.

(b) The documentation referred to in 3 above shall be retained and kept available for the competent national authorities for at least 10 years following the date of manufacture of the machinery or of the last unit produced, in the case of series manufacture.

(c) The documentation referred to in 3 above shall he drawn up in one of the official languages of the Communities, with the exception of the instructions for the machinery.

ANNEX VI
EC TYPE-EXAMINATION

For the purposes of this Annex, "machinery" means either "machinery" or "safety component" as defined in Article 1(2).

1. EC type-examination is the procedure by which a notified body ascertains and certifies that an example of machinery satisfies the provisions of this Directive which apply to it.

2. The application for EC type-examination shall be lodged by the manufacturer or by his authorized representative established in the Community, with a single notified body in respect of an example of the machinery. The application shall include:

- the name and address of the manufacturer or his authorized representative established in the Community and the place of manufacture of the machinery,
- a technical file comprising at least:
  - an overall drawing of the machinery together with drawings of the control circuits,
  - full detailed drawings, accompanied by any calculation notes, test results, etc., required to check the conformity of the machinery with the essential health and safety requirements,
  - a description of methods adopted to eliminate hazards presented by the machinery and a list of standards used,
  - a copy of the instructions for the machinery,
  - for series manufacture, the internal measures that will be implemented to ensure that the machinery remains in conformity with the provisions of the Directive.

It shall be accompanied by a machine representative of the production planned or, where appropriate, a statement of where the machine may be examined.

The documentation referred to above does not have to include detailed plans or any other specific information as regards the sub-assemblies used for the manufacture of the machinery unless a knowledge of them is essential for verification of conformity with the basic safety requirements.

3. The notified body shall carry out the EC type-examination in the manner described below:

- it shall examine the technical construction file to verify its appropriateness and the machine supplied or made available to it,
- during the examination of the machine, the body shall:

(a) ensure that it has been manufactured in conformity with the technical construction file and may safely be used under its intended working conditions;

(b) check that standards, if used, have been properly applied;

(c) perform appropriate examinations and tests to check that the machine complies with the essential health and safety requirements applicable to it.

4. If the example complies with the provisions applicable to it the body shall draw up an EC Type-examination certificate which shall be forwarded to the applicant. That certificate shall state the conclusions of the examination, indicate any conditions to which its issue may be subject and be accompanied by the descriptions and drawings necessary for identification of the approved example.

The Commission, the Member States and the other approved bodies may obtain a copy of the certificate and, on a reasoned request, a copy of the technical construction file and of the reports on the examinations and tests carried out.

5. The manufacturer or his authorized representative established in the Community shall inform the notified body of any modifications, even of a minor nature, which he has made or plans to make to the machine to which the example relates. The notified body shall examine those modifications and inform the manufacturer or his authorized representative established in the Community whether the EC type-examination certificate remains valid.

6. A body which refuses to issue an EC type-examination certificate shall so inform the other notified bodies. A body which withdraws an EC type-examination certificate shall so inform the Member State which notified it. The latter shall inform the other Member States and the Commission thereof, giving the reasons for the decision.

7. The files and correspondence referring to the EC type-examination procedures shall be drawn up in an official language of the Member State where the notified body is established or in a language acceptable to it.

ANNEX VII
MINIMUM CRITERIA TO BE TAKEN INTO ACCOUNT BY MEMBER STATES FOR THE NOTIFICATION OF BODIES

For the purposes of this Annex, "machinery" means either "machinery" or "safety component" as defined in Article 1(2).

1. The body, its director and the staff responsible for carrying out the verification tests shall not be the designer, manufacturer, supplier or installer of machinery which they inspect, nor the authorized representative of any of these parties. They shall not become either involved directly or as authorized representatives in the design, construction, marketing or maintenance of the machinery. This does not preclude the possibility of exchanges of technical information between the manufacturer and the body.

2. The body and its staff shall carry out the verification tests with the highest degree of professional integrity and technical competence and shall be free from all pressures and inducements, particularly financial, which might influence their judgement or the results of the inspection, especially from persons or groups of persons with an interest in the result of verifications.

3. The body shall have at its disposal the necessary staff and possess the necessary facilities to enable it to perform properly the administrative and technical tasks connected with verification; it shall also have access to the equipment required for special verification.

4. The staff responsible for inspection shall have:

- sound technical and professional training,
- satisfactory knowledge of the requirements of the tests they carry out and adequate experience of such tests,
- the ability to draw up the certificates, records and reports required to authenticate the performance of the tests.

5. The impartiality of inspection staff shall be guaranteed. Their remuneration shall not depend on the number of tests carried out or on the results of such tests.

6. The body shall take out liability insurance unless its liability is assumed by the State in accordance with national law, or the Member State itself is directly responsible for the tests.

7. The staff of the body shall be bound to observe professional secrecy with regard to all information gained in carrying out its tasks (except vis-à-vis the competent administrative authorities of the State in which its activities are carried out) under this Directive or any provision of national law giving effect to it.

ANNEX VIII
PART A REPEALED DIRECTIVES
(REFERRED TO BY ARTICLE 14)

Directive 89/392/EEC and its following amendments:

- Directive 91/368/EEC   only Article 1
- Directive 93/44/EEC
- Directive 93/68/EEC   only Article 6

Part B List of deadlines for transposition into and application in national law (referred to by Article 14)

Directive	Deadline for transposition	Date of application
Directive 89/392/EEC (OJ L 183, 29.6.1989, p. 9)	1 January 1992	Starting from 1 January 1993; for the products referred to by Directives 86/295/EEC, 86/296/EEC and 86/663/EEC: starting from 1 July 1995[1]
Directive 91/368/EC (OJ L 198, 22.7.1991, p. 16)	1 January 1992	Starting from 1 January 1993

Directive 93/44/EEC (OJ L 175, 19.7.1993, p. 12)	1 July 1994	• Starting from 1 January 1995[2]
		• Starting from 1 July 1994[2]
		  • Article 1(10), with the exception of points (a), (b) & (q)
		  • Article 1(11)(a) and (b)
		  • Article 1(12)(c), (d), (e) and (f)

Directive 93/68/EEC (OJ L 220, 30.8.1993, p. 1)	1 July 1994	Starting from 1 January 1995(3)

[1] For the period ending on 31 December 1994, the Member States should have authorized, except for he products referred to by Directives 86/295/EEC, 86/296/EEC and 86/663/EEC for which this period was ending on 31 December 1995, the placing on the market and putting into service of machinery which comply with the national provisions in force in their territories on 31 December 1992.

(2) For the period ending on 31 December 1996, the Member States shall allow the placing on the market and putting into service of machinery for the lifting or moving of persons as well as safety components which comply with the national provisions in force in their territories on 14 June 1993.

(3) Until 1 January 1997 Member States shall allow the placing on the market and putting into service of products which comply with the marking arrangements in force before 1 January 1995.

ANNEX IX
CORRELATION TABLE

Directive 89/392/EEC	This Directive
Article 1(1)	Article 1(1)
Article 1(2), first subparagraph	Article 1(2), point (a), first indent
Article 1(2), second subparagraph	Article 1(2), point (a), second indent
Article 1(2), third subparagraph	Article 1(2), point (a), third indent
Article 1(2), fourth subparagraph	Article 1(2), point (b)
Article 1(3)	Article 1(3)
Article 1(4)	Article 1(4)
Article 1(5)	Article 1(5)
Article 2	Article 2
Article 3	Article 3
Article 4	Article 4
Article 5	Article 5
Article 6	Article 6
Article 7	Article 7
Article 8(1)	Article 8(1)
Article 8(2)	Article 8(2)
Article 8(3)	Article 8(3)
Article 8(4)	Article 8(4)
Article 8(4a)	Article 8(5)
Article 8(5)	Article 8(6)
Article 8(6)	Article 8(7)
Article 8(7)	Article 8(8)
Article 9	Article 9
Article 10	Article 10
Article 11	Article 11
Article 12	Article 12
Article 13(1)	—
Article 13(2)	—
Article 13(3)	Article 13(1)
Article 13(4)	Article 13(2)
—	Article 14

Directive 89/392/EEC	This Directive
—	Article 15
—	Article 16
Annex I	Annex I
Annex II	Annex II
Annex III	Annex III
Annex IV	Annex IV
Annex V	Annex V
Annex VI	Annex VI
Annex VII	Annex VII
—	Annex VIII
—	Annex IX

Glossary

IIB declaration	Declaration that is not obligatory but is advisable to supply because a manufacturer can limit liability with it.
ad	ETSI date of adoption of the Standard.
Apparatus	All electrical and electronic appliances together with equipment and installations containing electrical and/or electronic components.
Appliance	Finished product with an essential function for the end user (single commercial unit).
Assembly on the company's own initiative	Assembly of products by a company according to the job
Authorized representative	One authorized to act as the representative of a manufacturer.
B1 standards	Standards that give a more detailed explanation of safety principles described generally in A standards-for example, EN 294 (Safety distances for preventing hazardous zones being reached by the upper limbs) and EN 349 (Minimum distances for preventing injury to human body parts).
B2 standards	Standards containing more detailed explanations of technical solutions of safety principles than described in A standards-for example, EN 574 (Two-handed operation) and EN418 (Emergency stop equipment).

Basic standards	Standards that provide a definition and description of the phenomenon, detailed test methods, and a description of the test facilities and measurement equipment to be used.
Bring into circulation	To make a product available for the first time, against payment or free of charge, under a Directive in the EC market for the purpose of distribution and/or use in the EC territory. (See also Launch on the market.)
CE	Conformité Européenne, compliance of a product with all the required New Approach Directives.
CEE-el	International Commission for the Inspection of Electrical Material.
CE Marking	Indication affixed to a product, standing for "Conformité Européene," which means "European Conformity." If products comply with the requirements of the New Approach Directives, then this is indicated by the letters "CE."
CEN	Comité Européen de Normalisation (European Commission for the Normalisation).
CENELEC	Comité Européen de Normalisation Electrotechnique (European Commission for Electrotechnic Normalisation).
Class 0	Protection consisting solely of basic insulation, no grounding.
Class 01	Protection consisting of basic insulation and grounding which, however, is not connected to the connecting lead (floating ground).
Class 2	Protection consisting of basic insulation and secondary or amplified insulation, known as "double-insulated."

Class 3	Protection obtained from safety extra low voltage (SELV) circuits, preventing higher voltages. A SELV circuit is one in which the electric voltage does not normally exceed a maximum of 42 V between the conductors and the ground, while the unloaded voltage can be a maximum of 50 V.
Class I	Protection consisting of basic insulation and connected grounding. The contactable metal components, which in the event of a defect could become live, are linked directly to the grounding of the lighting circuit.
Class I, II, or III	The equipment insulation class. Class I equipment uses basic insulation and is connected to a ground wire. Class II equipment has double insulation, while Class III equipment is provided with a safe, very low voltage, from a safety extra low voltage (SELV) circuit.
Compatibility	Ability of a device, unit of equipment, or system to function satisfactorily in its electromagnetic environment without introducing intolerable electromagnetic disturbances to anything in that environment.
Competent Body	An external certification body for the EMC Directive designated by the national governments, or any body that meets the criteria listed in Annex II and is recognized as such.
Component	Any part which is incorporated in an appliance and which is not an appliance in itself with an essential function for the end user.
Composition of Machines	An installation or production line which consists of two or more machines.

Creepage distances and clearances	Creepage is the shortest distance between two conducting components over a surface while clearance is the shortest distance between two conducting components through the air.
CTR	Common Technical Regulations. Standards for telecommunications equipment.
Damage	Physical injury and/or damage to health or objects.
dea	CENELEC deadline for reply of national Standards Bodies and National Committees on proposed Standards.
Decisions	Determinations that are binding in all their parts on those to whom they are expressly directed; this means that a Decision can apply to both Member States and private individuals.
DECT	Mobile telephone.
Designated Laboratory	An external laboratory designated by the national Governments.
DIN	Deutsche Institut für Normung (German Normalisation Institute).
Direct applicability of a Regulation	Property whereby a Regulation takes effect without first having to be transposed into national legislation.
Direct function	Any function of a component itself that satisfies the intended use as described in the instructions for the end user supplied by the manufacturer.
Directive	A European law that is legally binding for every Member State and which is above the laws of the individual Member States. A Directive is not aimed directly at the citizens or companies in a Member State.
doa	CENELEC latest date of Announcement of the Standard at a national level.

dop	CENELEC latest date of Publication of the Standard by publication at a national level.
dor	CENELEC date of Ratification by the Technical Board of CENELEC.
dow	CENELEC latest date of Withdrawal of conflicting National Standards.
EC	European Community.
EC Declaration of Conformity	A document in which a manufacturer in the EEA, an authorized agent of the manufacturer domiciled in the EEA, or an importer officially declares that a product complies with all the essential requirements of a Directive.
EC Declaration of Conformity IIA	Document intended for machinery or machine accessories.
EC Declaration of Conformity IIB	A statement that machinery must not be put into service until the machinery into which it is to be incorporated has been declared in conformity with the provisions of the Directive.
EC Declaration of Conformity IIC	Declaration intended for safety components.
ECSC	European Coal and Steel Community.
EC type examination	Examination involving the inspection of a representative example (type) of the series in question by, or on behalf of, an external inspection organisation or Notified Body within the EEA.
EC type-examination certificate:	Document in which a notified body referred to in Article 10 (6) certifies that the type of equipment examined complies with the provisions of the Directive concerned.
EEA	European Economic Area. The fifteen countries of the European Union as well as Iceland, Norway, and Liechtenstein belong to the EEA.

EEC	European Economic Community.
Electric torch	Flashlight.
Electromagnetic disturbance	Any electromagnetic phenomenon that may degrade the performance of a device, unit of equipment, or system.
EMC	Electromagnetic compatibility.
EMF	Electro Motor Force.
EN	An accepted European Standard.
ESD	Electrostatic discharge.
ETS	An accepted European Telecommunication Standard.
ETSI	European Telecommunication Standardization Institute.
EU	European Union.
Euratom	European Atomic Energy Commission.
EUT	Equipment under test.
FOPS	Falling object protective structure.
General Standards	Standards that contain the limit values that must be generally applied for those products to which the EMC phenomena are considered to be relevant.
General Tenor	Quality of a Regulation whereby in principle it is applicable to an indeterminate number of cases and persons.
GS	Geprüfte Sicherheit. German quality and safety mark.
GSM	Mobile telephone.
Hand-held equipment	A device intended to be held in the hand during normal use.
Harmonized European Standards (EN)	Standards published in the Official Journal of the European Communities that are intended to harmonize the legislation of the Member States.

Hazardous event	An event which can cause damage.
Hd	A harmonization document.
IEC	International Electrotechnical Commission.
IEC-CISPR	Comité International Spécial de Perturbations RadioÈlectrique (Special International Commission for Radioelectric Perturbations).
I-ETS	An accepted Interim European Telecommunication Standard.
Immunity	Ability of a device, unit of equipment, or system to perform without degradation of quality in the presence of an electromagnetic disturbance.
Inspectorate of Goods	Market surveillance body.
Installation	A combination of various appliances and/or systems put together at a specific location with a special purpose that should not be launched on the market as a single commercial unit.
Interchangeable equipment	Tool or component with which the original function of a machine can be changed. The feature of such equipment is that it can be connected and disconnected by the user.
International standards (IEC/CEE)	Standards that may be applied if no harmonized standards have yet been established.
ISM	Industrial, scientific, and/or medical.
ISO	International Standardization Organization.
ITE	Information technology equipment.
ITU	International Telecommunications Union.
KEMA KEUR	Dutch certification body for quality and safety.
KOMO	Dutch certification body for quality.
Labor Inspectorate	Market surveillance body.

Launch on the market	To make a product available for the first time, against payment or free of charge, under a Directive in the EC market for the purpose of distribution and/or use in the EC territory. (See also Launch on the market.)
Leakage current	The current that during normal operations runs from the power conductors to ground.
Liability	Legal duty to provide compensation for damages caused by a defect in a product.
Machinery	An assembly of linked parts or components, at least one of which moves, with the appropriate actuators, control and power circuits, etc., joined together for a specific application, in particular for the processing, treatment, moving, or packaging of a material.
Module C	Conformity to the type.
Module D	Production quality assurance.
Module E	Product quality assurance.
Module F	Product verification.
Module G	Unit verification.
Module H	Full quality assurance, according to EN 29001.
Movable equipment	A portable device weighing a maximum of 18 kg or equipment on casters that the user can move around.
MRA	Mutual Recognition Agreement.
National standards	Standards (e.g., NEN, DIN, NF, BS) of a country to which export is to take place that may be applied if no international standards exist.
Negative default	Damage as a result of nonobservance or late observance of an agreement as well as the damage arising when a penalty, com-

	pensation, guarantee, indemnification, or other similar condition is not met.
New Approach	European Directives per product group without further explanation of detailed technical specifications. Imposes essential requirements in the field of safety, health, environment, and consumer protection. These requirements have been elaborated in the harmonized standards.
New Approach Directives	The Directives under the New Approach, imposing essential requirements in the field of safety, health, environment, and consumer protection.
Non-automatic weighing machines	Weighing instruments in which the intervention of an operator is required for the weighing activity.
Normal load	Normal operating status of a device, with loads applied to every possible connection or output.
Notified Body	An external certification body designated by the national governments.
OATS	Open-area test site.
Old Approach	Former approach under which a Directive was formulated for each product in which all the technical specifications and standards were formulated. The requirements with which a product had to comply were established right down to the nearest detail.
PC	Personal computer.
Power distribution system	Choice of electrical current system used for power supply. Frequently occurring variants include the TN, TT, and IT systems. The first letter indicates whether the supply source has a ground connection: "T" is connected, "I" is insulated. The second letter indicates how the installation is connected

to ground: in the case of an "N," the frames are connected to the neutral point of the power supply unit using a conductor; in the case of a "T," the frames are connected locally to ground. So for a TN system, the ground connection is linked to the power-supply ground connection via the lead wires while for a TT and an IT system, the ground connection is made separately from the device.

prEN — A draft European Standard.

prENV — A draft Provisional European Standard.

prETS — A draft European Telecommunication Standard.

prI-ETS — A draft Interim European Telecommunication Standard.

Private label — A commercial company's own trade name applied to purchased products.

Product Group Standards — Also called Product Family Standards; standards intended for a group of related products.

Product Standards — The same as product family standards except that the product standard, in contrast to the product family standard, is intended for one specific type of product.

Rated voltage, current, frequency — Nominal supply voltage, current and frequency.

Regulations — Rules that have a general tenor and all of whose components are binding and are directly applicable to all Member States.

Remaining risk — The risk which cannot be eliminated by a safety measure.

Risk assessment — Process for evaluating risk, the basis for a safe product and also the starting point for the user's manual and the Technical Construction File.

Risk category A	Trapping or injury of fingers between blades.
Risk category B	Hair or loose clothing getting into blades.
Risk category C	Cutting through the electric cable during trimming.
Risk category D	Use in a humid environment (risk of short circuit).
Risk category E	Other inappropriate use.
ROPS	Rollover protective structure.
Safety component	Equipment used by manufacturers of machines which are in principle unsafe in order to make the machines comply with the requirements of the Directive and to increase their safety level.
Safety measure	A means which eliminates a danger or reduces a risk.
Single commercial unit	Functional unit which is launched separately on the market and is very easy for the consumer to install.
Standard (EN)	A formulated criterion for attaining unity in a field in which diversity is inefficient or unnecessary.
Standardization	Process in which rules (standards) are drawn up in order to create order or unity wherever diversity is undesirable and/or unnecessary.
Standards	See Basic standards, General standards, and Product (family) standards.
Stock	A quantity of products that is present in the distribution chain for sale.
System	A combination of various appliances with the intention of performing a specific task and launched on the market as a single commercial unit.
TCF	Technical Construction File.

Technical Construction File — Document that contains the technical basis and that must show that all the criteria set out in the Directive have been met.

Transaction damage — Damage as a result of replacement, improvement, or repair of supplied products as well as damage resulting from inability or insufficient ability to use the products.

Type A standards — Standards covering general safety aspects that can be applied to all machines. They form the basic tools for every designer/constructor.

Type B standards — Standards that discuss specific technical safety aspects and safety provisions and can be implemented universally for all machines.

Type C standards — Standards that set safety specifications for certain (groups of) machines.

User's manual — An essential adjunct to every machine, containing user's instructions, safety regulations, and other regulations; can be legally defined as a set of (obligatory) rules of conduct for the user.

Quotation of Sources

Coenraads, ing. J.D., T.E.T. Koning, ir. J.G. van Hezewijk, ing. H. Tander e.a., Handbook CE Marking in electrical and mechanical engineering, Kluwer Bedrijfsinformatie

Index

A

Active implantable medical devices Directive, 32–33
ad, defined, 349
Agri–foodstuffs machinery
  in Machinery Directive 89/392/EEC, 174–175
  in Machinery Directive 98/37/EC, 309–310
Allied Quality Assurance Publications (AQAP) certificate, 24–25
American manufacturer, European Union (EU) and, 27–32
Apparatus, defined, 349
Appliance, defined, 349
Appliances burning gaseous fuels Directive, 35
AQAP (Allied Quality Assurance Publications) certificate, 24–25
Assembly on the company's own initiative, defined, 349
Audits by national organizations, 23–24
Austria addresses with regard to the Machinery Directive, 264
Authorized representative, defined, 349

B

B1 standards, defined, 349
B2 standards, defined, 349
Basic standards, defined, 350
Belgium addresses with regard to the Machinery Directive, 255
Bring into circulation, defined, 350
Building materials Directive, 40

C

Cableway installations for personnel transport Directive, 42–43
CE (Conformité Européenne), 4
  defined, 350
CE conformity marking in Machinery Directive 98/37/EC, 337–338
CEE–el, defined, 350
CE Marking, 25, 27
  affixing, 96, 121–122
  defined, 350
  ISO 9000 and, 27
  liability and, 127–130
  in Machinery Directive 93/68/EEC, 44–45
  in Machinery Directive 98/37/EC, 285

CE Marking *(continued)*
  procedure for, under Machinery
    Directive, 75–96
  responsibilities for, 61–67
CEN (Comité Européen de
  Normalisation), 13, 18–19
CENELEC (Comité Européen de
  Normalisation
  Electrotechnique), 13, 18–19
Certification procedure for
  Machinery Directive
  89/392/EEC, 150–152
Check list(s), 245–252
  for end producer against liability,
    134–138
  environment, 248
  other resources, 251–252
  person, 245–246
  process tree, 248–251
  product, 246–247
Class 0, defined, 350
Class 01, defined, 350
Class 2, defined, 350
Class 3, defined, 351
Class I, defined, 351
Class I, II, or III, defined, 351
Comité Européen de Normalisation
  (CEN), 13, 18–19
Comité Européen de Normalisation
  Electrotechnique (CENELEC),
  13, 18–19
Compatibility, defined, 351
Competent body, defined, 351
Complaints, handling, 131
Component, defined, 351
Composition of machines, 51
  defined, 351
Conditions of use, identification of,
  110–112
Conformité Européenne (CE), 4
  defined, 350

Conformity–assessment procedures,
  19–24
  in Machinery Directive 98/37/EC,
    282–285
Contracts, adapting, 133
Contractual liability, 129
Control devices
  in Machinery Directive
    89/392/EEC, 157
  in Machinery Directive 98/37/EC,
    291–292
Controls
  in Machinery Directive
    91/368/EEC, 194–196
  in Machinery Directive 98/37/EC,
    290–295, 314–316
Creepage distances and clearances,
  defined, 352
Criminal liability, 130
CTR, defined, 352

D

Damage, 77
  defined, 352
dea, defined, 352
Decisions, 8–9
  defined, 352
DECT, defined, 352
Denmark addresses with regard to
  the Machinery Directive, 255
Designated Laboratory, defined, 352
Design errors, 126
DIN, defined, 352
Direct applicability of a Regulation,
  defined, 352
Direct function, defined, 352
Directive(s), 8
  defined, 352
doa, defined, 352
Domestic electric refrigerators and
  deep freezers Directive, 39

dop, defined, 353
dor, defined, 353
dow, defined, 353

E

EC (European Community), 3
EC Declaration of Conformity, 28, 120, 121
  defined, 353
  in Machinery Directive 89/392/EEC, 179–181
  in Machinery Directive 98/37/EC, 340–341
EC Declaration of Conformity IIA, 92–93, 106
  defined, 353
EC Declaration of Conformity IIB, 93–95, 107
  defined, 353
EC Declaration of Conformity IIC, 95
  defined, 353
EC mark in Machinery Directive 89/392/EEC, 152, 178
ECSC (European Coal and Steel Community), 3
EC Type Declaration, 91
EC type–examination
  defined, 353
  in Machinery Directive 89/392/EEC, 181–183
  in Machinery Directive 98/37/EC, 341–343
EC type–examination certificate, defined, 353
EEA (European Economic Area), ix
  defined, 353
EEC (European Economic Community), 3
Efficiency requirements for new hot–water boilers fired with liquid or gaseous fuels Directive, 38–39
Electric torch, defined, 354
Electromagnetic Compatibility Directive, 34–35
Electromagnetic disturbance, defined, 354
EMC, defined, 354
EMF, defined, 354
EN 292-1, 57
EN 292-2, 57
EN 294, 59–60
EN 349, 60
EN 414, 57–58
EN 1050, 58
EN 60204-1, 59
EN (European Norm), 18
"EN," prefix, 56
EN standards, list of, 231–242
ENV 1070, 58
Environmental requirements in user's manual, 90–91
Environment check list, 248
Equipment and protective systems Directive, 33–34
ESD, defined, 354
ETS, defined, 354
ETSI (European Telecommunication Standardization Institute), 13, 19
EU, *see* European Union
European Atomic Energy Commission (Euratom), 3
European Coal and Steel Community (ECSC), 3
European Commission, 11
European Community, *see* EC *entries*
European Conformity, 4
European Economic Area (EEA), ix
  defined, 353

European Economic Community
    (EEC), 3
European Norm (EN), 18
European Telecommunication
    Standardization Institute
    (ETSI), 13, 19
European Union (EU), 3
    American manufacturer and, 27–32
    countries of, 4
    history of, 3–6
    rules and regulations of, 6–9
EUT, defined, 354
Explosives for civil use Directive, 41

F
Finland addresses with regard to the
    Machinery Directive, 258–259
FOPS, defined, 354
Foreseeable abnormal conditions of
    use, 111
France addresses with regard to the
    Machinery Directive, 259–261
Free movement of goods, 4

G
General Standards, defined, 354
General Tenor, defined, 354
Germany addresses with regard to
    the Machinery Directive,
    255–258
Glossary, 349–360
Goods, free movement of, 4
Grey areas, existence of, 13–15
GS, defined, 354
GSM, defined, 354
Guards and protection devices,
    required characteristics of
    in Machinery Directive
        89/392/EEC, 164–166
    in Machinery Directive 98/37/EC,
        298–300

H
Hand–held equipment, defined,
    354
Harmonization, technical, 17–19
Harmonized European Standards
    (EN), defined, 354
Hazardous event, 77
    defined, 355
Hazards, protection against
    in Machinery Directive
        89/392/EEC, 166–169
    in Machinery Directive
        91/368/EEC, 199–200
    in Machinery Directive 98/37/EC,
        300–303, 319–320
Hd, defined, 355
Health and safety requirements
    in Machinery Directive
        89/392/EEC, 153–176
    in Machinery Directive
        91/368/EEC, 191–212
    in Machinery Directive 93/44/EEC,
        220–228
    in Machinery Directive 98/37/EC,
        287–335
High–speed trains Directive, 42

I
IEC, defined, 355
IEC–CISPR, defined, 355
I–ETS, defined, 355
IIB declaration, defined, 349
Immunity, defined, 355
Indications
    in Machinery Directive
        91/368/EEC, 200–202
    in Machinery Directive 98/37/EC,
        320–322
Indicators
    in Machinery Directive
        89/392/EEC, 170–173

in Machinery Directive 98/37/EC, 305–309
Inspection of products supplied, 132–133
Inspectorate of Goods, defined, 355
Installation, defined, 355
Instruction errors, 126
Insurance against liability, 129–130
Interchangeable equipment, 50–51, 108
 defined, 355
Internal manufacturing inspection, 19
International standards (IEC/CEE), defined, 355
In vitro diagnostics Directive, 43–44
Ireland addresses with regard to the Machinery Directive, 261
ISM, defined, 355
ISO 9000, 25
 CE Marking and, 27
ISO, defined, 355
Italy addresses with regard to the Machinery Directive, 261–262
ITE, defined, 355
ITU, defined, 355

J
Japan, 24

K
KEMA KEUR, defined, 355
KOMO, defined, 355

L
Labor Inspectorate, defined, 355
Laser machines, 105
Launch on the market, defined, 356
Leakage current, defined, 356
Liability, 125–138

approach in event of damage and claims, 134–138
CE Marking and, 127–130
contractual, 129
criminal, 130
defined, 125, 356
exclusion of, 127
insurance against, 129–130
no complete exclusion of, 127–128
preventive measures against, 130–133
proving, 126–127
reasons for, 126
Lifts Directive, 42
Low–Voltage Directive, 35–36

M
Machine parts, 51–52
Machinery, defined, 356
Machinery Directive, 36, 49–72, ix
 addresses with regard to, 255–267
 Annex IV, 54
 area of application, 49–52
 CE Marking under, 75–96
 exceptions to, 52–54
 formalities of, 99
 for machines with increased risk, 54–55
 most frequently asked questions on, 99–109
 practical examples of, 99–122
 relations with other Directives, 55–56
 risk assessment and, 99–100, 112–113
Machinery Directive 89/392/EEC
 Annex I, 153–176
 Annex II, 176–177
 Annex IV, 178–179
 Annex V, 179–181
 Annex VI, 181–183

Machinery Directive 89/392/EEC
(*continued*)
  Annex VII, 183–184
  application of, 145–150
  certification procedure, 150–152
  correlation table with Machinery Directive 98/37/EC, 346–347
  EC mark in, 152, 178
  final provisions, 152–153
  health and safety requirements in, 153–176
  text of, 141–184
Machinery Directive 91/368/EEC
  Annex I, 191–212
  Annex II, 212–213
  text of, 185–213
Machinery Directive 93/44/EEC
  Annex I, 220–228
  text of, 214–228
Machinery Directive 98/37/EC, x
  Annex I, 287–335
  Annex II, 335–337
  Annex III, 337–338
  Annex IV, 338–339
  Annex V, 340–341
  Annex VI, 341–343
  Annex VII, 343–344
  Annex VIII, 345–346
  Annex IX, 346–347
  CE conformity marking in, 337–338
  CE Marking in, 285
  conformity assessment procedures in, 282–285
  correlation table with Machinery Directive 89/392/EEC, 346–347
  EC Declaration of Conformity in, 340–341
  EC type–examination in, 341–343
  final provisions, 285–287
  health and safety requirements in, 287–335
  repealed Directives, 345–346
  scope, placing on market and freedom of movement, 276–282
  text of, 271–347
Machinery for working wood and analogous materials
  in Machinery Directive 89/392/EEC, 176
  in Machinery Directive 98/37/EC, 311
Machines
  hiring of, 70–71, 108–109
  identification of, 109, 110
  installation of, 68–70
  sale of, 71–72
  in stock, 64
Maintenance
  in Machinery Directive 89/392/EEC, 169–170
  in Machinery Directive 98/37/EC, 304–305
Manufacturer
  American, European Union (EU) and, 27–32
  assembler of semi–finished products, 67–68
  authorized representative of, 65–66
  of complete machines, 63–64
  of machine parts and semi–finished products, 65
Measuring equipment Directive, 42
Mechanical hazards, protection against
  in Machinery Directive 89/392/EEC, 161–164
  in Machinery Directive 91/368/EEC, 196–199

in Machinery Directive 98/37/EC, 295–298, 316–319, 323–327
Medical devices Directive, 37
Member States, 3
differences between, 13
Minimum criteria to be taken into account by Member States for notification of bodies
in Machinery Directive 89/392/EEC, 183–184
in Machinery Directive 98/37/EC, 343–344
Module A, 20, 21
Module B, 21–22
Module C, 22
defined, 356
Module D, 22
defined, 356
Module E, 22
defined, 356
Module F, 22–23
defined, 356
Module G, 23
defined, 356
Module H, 20, 23
defined, 356
Movable equipment, defined, 356
Mutual recognition agreement (MRA), 31–32

N

National organizations, audits by, 23–24
National standards, defined, 356
Negative default, defined, 356–357
"NEN," prefix, 56
Netherlands addresses with regard to the Machinery Directive, 263
New Approach, defined, 357

New Approach Directives, 4–5, 9, 10, 11–15
advantages and limitations of, 12–15
defined, 357
disadvantages of, 13–15
Old Approach Directives versus, 12
overview of, 32, 33
Non–automatic weighing machines, defined, 357
Non–automatic weighing machines Directive, 37
Normal load, defined, 357
Norway addresses with regard to the Machinery Directive, 263
Notified Body, 19, 103
defined, 357

O

OATS, defined, 357
Official Journal of the EU (OJ/EC), 11
Old Approach, defined, 357
Old Approach Directives, New Approach Directives versus, 12

P

PC, defined, 357
Personal protective equipment Directive, 37–38
Person check list, 245–246
Pleasure craft Directive, 40–41
Portable hand–held and/or hand–guided machinery
in Machinery Directive 89/392/EEC, 175–176
in Machinery Directive 98/37/EC, 310–311
Portugal addresses with regard to the Machinery Directive, 264

Power distribution system, defined, 357–358
"pr," prefix, 56
Precious metals Directive, 43
prEN, defined, 358
prENV, defined, 358
Pressure equipment Directive, 41
prETS, defined, 358
prI–ETS, defined, 358
Private Label, 68
  defined, 358
Process tree check list, 248–251
Product check list, 246–247
Product Group Standards, defined, 358
Production errors, 126
Product Standards, defined, 358

Q
Quality, safety and, 133
Quotation of sources, 361

R
Rated voltage, current, frequency, defined, 358
Recommendations and Advice, 9
Regulations, 6–8
  defined, 358
  hierarchy of, 7
Remaining risk, 77
  defined, 358
Resources check list, 251–252
Responsibilities for CE Marking, 61–67
Risk assessment, 75–80, 99–100
  defined, 358
  Machinery Directive and, 112–113
  practical example of, 109–121
  risk inventory versus, 107
  risk reduction in, 116–119
  specific requirements for, 113–114

  specific standards for, 114
  user's manual and, 100
Risk category A, defined, 359
Risk category B, defined, 359
Risk category C, defined, 359
Risk category D, defined, 359
Risk category E, defined, 359
Risk evaluation, 116
Risk identification, 109
Risk inventory, risk assessment versus, 107
Risk reduction, 79–80, 116–119
Risks
  assessment of, 78–79
  evaluation of, 79
  identification of, 77–78
ROPS, defined, 359
Rules, hierarchy of, 7

S
Safety, quality and, 133
Safety component(s), 50, 51
  defined, 359
Safety evaluation, 80, 119–121
Safety level, 102
Safety measure, 77
  defined, 359
Safety of toys Directive, 39
Self–certification, 21
Semi–finished products, 51–52
Simple pressure vessels Directive, 34
Single commercial unit, defined, 359
Sources, quotation of, 361
Spain addresses with regard to the Machinery Directive, 264–266
Standard (EN), defined, 359
Standardization
  defined, 359
  term, 16–17
Standards, 15–19, 56–61
  defined, 15–16, 359

Step plan, 77
  against liability, 130–133
Stock, defined, 359
Stopping device
  in Machinery Directive 89/392/EEC, 158–159
  in Machinery Directive 98/37/EC, 292–293
Sweden addresses with regard to the Machinery Directive, 267
Switzerland, 6
System, defined, 359

T

Technical Construction File (TCF), 28, 80–81, 106
Telecommunications terminal equipment Directive, 38
Trade barriers, 10
  removal of, 12–13
Transaction damage, defined, 360
Translations of user's manual, 100–101, 102, 104, 105
Type A standards, 56–58
  defined, 360
Type B standards, 58–60
  defined, 360
Type C standards, 60–61
  defined, 360

U

Unanimity requirement, replacement of, 10–11
United Kingdom addresses with regard to the Machinery Directive, 266–267

Users, identification of, 109–110
User safety, 28
User's manual, 81–91
  contents of, 101
  copyright of, 107
  defined, 360
  dismantling hazards in, 88
  environmental requirements in, 90–91
  installation requirements in, 88
  instructions for use in, 89
  language of, 82
  maintenance and repair provisions in, 89–90
  methods and tools in, 83–84
  retention of, 103–104
  risk assessment and, 100
  safety aspects in, 85–87
  structure of, 82–83
  technical data in, 84
  translations of, 100–101, 102, 104, 105
  transportation provisions in, 87–88

W

White Paper on Completion of the Internal Market, 10
Work stations
  in Machinery Directive 89/392/EEC, 192–194
  in Machinery Directive 98/37/EC, 312–314